Mid-Size Track Plans for
Realistic Layouts

BERNARD KEMPINSKI

KALMBACH
BOOKS

Acknowledgements

I would like to thank the following individuals for their help in preparing this book. First, Mark Thompson at Kalmbach Publishing for his inspiration and confidence in me. Jeff Wilson was extremely helpful in finding photos in Kalmbach's David P. Morgan Library and elsewhere.

Paul Dolkos, Tom Flagg, Mac Beard of the Chesapeake & Ohio Historical Society, Gene Huddleston, James Eudaly, Kevin Eudaly, Gary Kohler, Mike Danneman, John Leopard, Ron Burkhardt, Dan Mackey, Tim Darnell, Tony Koester, and Don McFall of the Western Maryland Railway Historical Society all provided photos for various chapters. Robert Kempinski, Jeff Kohler, and D.C. Cebula provided review comments for selected chapters.

My wife, Alicia, is due special recognition for her unfailing support, encouragement, and understanding during the preparation of this project.

Finally, I want to dedicate this book to the men and women of the armed services whose sacrifice and devotion to duty make hobbies like model railroading possible. Thank you.

Bernard Kempinski

About the author

Bernard Kempinski is a freelance writer who has written more than 40 magazine articles on model railroading, many of them on layout planning. He is an active model railroader and has built many models on commission, including a 1950s steel mill and a paper mill featured in recent Walthers catalogs.

A former U.S. Army captain, Bernard works as an analyst for the Institute for Defense Analysis in Alexandria, Va.

Printed in the United States of America
12 11 10 09 08   1 2 3 4 5

Secure online ordering available: www.KalmbachBooks.com

Publisher's Cataloging-In-Publication Data
(Prepared by The Donohue Group, Inc.)

Kempinski, Bernard.
    Mid-size track plans for realistic layouts / Bernard Kempinski

    p. : ill., maps, plans ; cm.

    ISBN: 978-0-89024-704-4

1. Railroad tracks--Models--Design and construction. 2. Railroads--Models.
3. Railroads--Models--Design and construction. 4. Railroads--United States--History. I. Title. II. Title: Realistic layouts

TF197 .K457 2008
625.1/9

Contents

Introduction

This book presents a set of well-researched mid-size model railroad track plans that should appeal to the typical, busy model railroad hobbyist. The plans are carefully designed to fit typical room, basement, and home garage spaces of 400 square feet or less. All of the track plans are shown in multiple scales to appeal to the widest variety of modelers, while simultaneously showcasing the advantages of that scale. Each of the 10 chapters covers a different prototype railroad. The track plans range from an 8 x 10-foot switching district to a 20 x 20-foot G scale layout designed for a garage. The plans cover many eras and many areas of the country as shown by the map at right.

Although the plans have a strong prototype basis, they do not follow their prototypes slavishly if the design can be improved. For example, industries and facilities that are spaced apart in the prototype may be condensed into one location or town.

The mid-size layout space itself tends to restrict the subjects that can be modeled. It doesn't make a lot of sense to try to fit the Pennsylvania Railroad's four-track main line around Horseshoe Curve or Chicago's Union Station in a medium-sized layout space, especially if the plan is meant to be operated. Instead, the plans in this book focus on mineral roads, branch lines, and switching districts, all subjects of limited scope, but not limited interest. By reducing the scope of the layout, the plan can simulate the real line with higher fidelity. While it's impractical to model heavy mainline action in a small- to medium-sized room, chapter 4 shows how double-decking can help if the subject is carefully selected.

Guidelines

The plans consider several design guidelines. First, there is minimal hidden track. Over the years, I have learned that extensive hidden track is rarely worth the increased trouble and aggravation. Some hidden track is acceptable as long as provisions are made for maintenance and access.

Second, the plans are "sincere," meaning tracks don't cross through the same scene twice unless they do so on the prototype. Real railroads are meant to deliver passengers and cargo from point to point. That is the illusion we want to achieve on a realistic model railroad, and that illusion can be shattered if the tracks crisscross the scene. Sincere layout designs also have the advantage of being easier to build. As such, they can be can be put into operation more quickly. If changes are needed, it's easier to modify a sincere plan than a complex spaghetti bowl, where every square inch of layout has been occupied with track.

Next, the plans feature as little selective compression as possible. If a plan models a mainline run, it focuses on only a short portion of it, with that portion modeled without crowding. Staging represents the rest.

A high scenery-to-track ratio adds realism to most plans. This conveys a feeling of distance by implying that the railroad is running through the countryside. This is where smaller scales excel. However, models in larger scales tend to run better, especially steam locomotives and switchers.

Studying the plans in this book, where designs for a particular railroad are presented in two different scales, should give you some insight to the trade-offs involved. If you're undecided on a modeling scale, seeing these plans may help you select one even if you don't plan to model the prototypes covered in this book.

Curves on each plan are as broad as possible, with No. 6 or 8 turnouts in most cases. This makes layouts look and operate better.

All the plans are designed for walk-around operation, preferably with wireless Digital Command Control. Once you've used wireless throttles, you'll probably agree that it is the way to go.

The easiest-to-build feature of any model railroad is the aisle. Once you begin operating a layout, you'll find that the convenience of generous aisles far outweigh the benefits of including an extra design feature such as an additional peninsula or turn-back loop. In these designs, I strived for aisles a minimum of three feet wide. In cases where an aisle is narrower, there are "staging areas" on either side of the narrow portion where operators can gather without crowding.

The plans avoid duckunders, especially those that require an operator to duck under benchwork while running a train. Although several plans require ducking under to enter a layout, they are designed either as taller "nod-unders" or include the potential for removable sections.

Prototype scenes

All of the plans include signature scenes that convey to the visitor or operator the theme of the layout. Scenes like the distinctive rock outcroppings of Castle Gate in Chapter 7 or the Wiscasset waterfront scene in Chapter 10 immediately set the stage for the railroad action that will follow.

Care was taken to ensure that the designs will fit in the stated space. Computer-aided design tools took the guesswork out of drawing standard track elements and prevented me from "cheating." All curves include easements for best appearance and operation. The plans in each chapter should all translate well to physical layouts.

The plans do not include detailed construction drawings. However, there are

Locations of prototype railroads modeled

some construction techniques that I have found useful over the 17 years I've been building layouts and modules.

I prefer open-grid benchwork for its sturdiness and portability. According to the U.S. Census, approximately 15 percent of Americans move each year, so designing a layout for portability is prudent. I have also found that open-grid benchwork is easy to convert to new designs when you decide to try something new. I prefer using cookie-cutter roadbed. I find it easier to visualize a track plan when laying out the design on a sheet of clean plywood. It is easy to mock up structures and scenery features on the plywood.

For details on building benchwork, see books on the topic such as *Basic Model Railroad Benchwork* or *How to Build Model Railroad Benchwork* (both published by Kalmbach).

I've become a believer in using .060" styrene plastic for backdrops. I get it at my local plastics supplier in 4 x 8-foot sheets. It's easy to cut with a straightedge and hobby knife and it's sturdy and dimensionally stable. Making seams is just like build-ing any styrene model. I use a large brush to apply copious amounts of styrene cement using a strip of scrap styrene behind the joint to reinforce it. Then I use model putty to fill in the seam. I have never had a seam crack. It's also flexible in case changes are needed. Once primed with gesso, it takes paint very well.

In line with the portability theme, I recommend using modular wiring. Modular standards have been thoroughly proven over the years for robust performance. Doing this means layout sections can be unplugged and isolated very easily, making it easier to find wiring problems.

Further research

All the track plans in this book are based on actual railroads. Researching the information to design a realistic railroad can be a hobby by itself. The place to begin is the historical society for the railroad you're interested in. Most have Web sites.

The Library of Congress and the U.S. National Archives have a wealth of rail-road-related information. The Historical American Engineering Record as well as Interstate Commerce Commission valua-tion records are particularly useful. If you don't live in the Washington, D.C. area, you can still research their holdings by visiting their Web sites (a good place to start is www.loc.gov). Both agencies regu-larly add scanned materials.

The recent advent of searchable map-ping software has made Web sites like Google Earth, Terraserver, and for con-temporary railroads, Live Local Birds Eye View indispensable. The Terraserver is especially useful as its maps are based on U.S. Geological Survey data and are in the public domain.

For more examples of layout designs, philosophy, and case studies, I recommend the annual magazine *Model Railroad Planning*, published by Kalmbach. It is packed with great model railroad design information. The National Model Rail-road Association's Layout Design Special Interest Group is another resource worth checking out. The group publishes a well-done journal on a periodic basis. It covers layout design subjects in more detail than commercial magazines are able to do.

CHAPTER 1

White Pass & Yukon

Military supplies head north on the WP&Y in 1942. Because of the railroad's strategic location, World War II brought a tremendous increase in traffic on the line. *U.S. Army*

The narrow gauge White Pass & Yukon was built through some of the roughest territory in North America. Although the railroad owes its history to gold and greed, it was its strategic location at the outbreak of World War II that caused its greatest traffic growth. Basing a layout on the WP&Y during that period provides a chance to model a narrow-gauge railroad with dramatic scenery that boasted more traffic than some standard gauge main lines of its day.

White Pass & Yukon

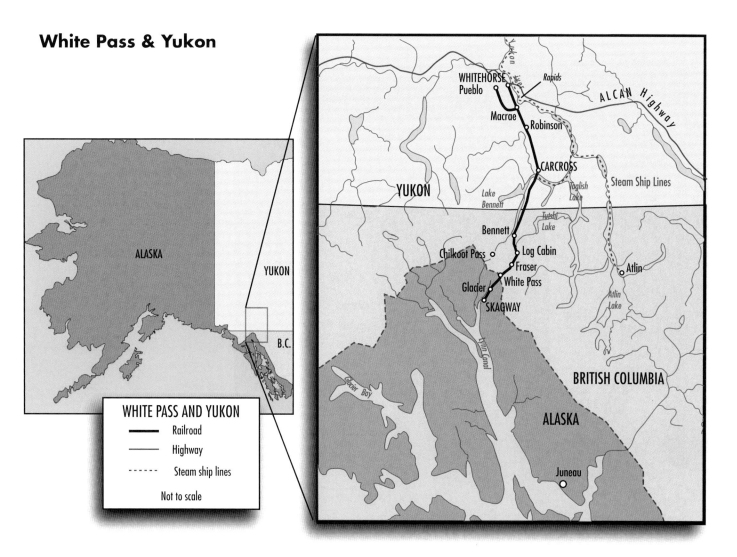

History

The extent that man will go to strike gold is beyond modern comprehension. Nowhere was this more obvious than the Yukon Gold Rush of 1898. Prospectors on the Klondike River discovered gold in 1897. However, due to the remoteness of the area, word did not reach the U.S. until a year later when a steamer reached Seattle with over $1 million in gold dust on board.

Then the stampede began. More than 100,000 people stricken with "gold fever" ventured north. Through incredibly harsh conditions, including the cold Arctic winter and rugged terrain, they rushed to the Yukon to stake their claims. In the end, less than half of these adventurers made it to their destinations.

Three routes were available to the stampeders. The overland route through Canada was the most difficult, while steaming up the Yukon River was easiest. Because the river wasn't navigable in the winter, many elected to take the third route, via the inland passage from Seattle or Vancouver to Dyea (Skagway), Alaska. There, they went by foot over the Chilkoot Trail. The famous "golden stair-case," 1,500 steps carved out of snow and ice, was at the summit of this trail. Too steep for packhorses, stampeders had to cache their goods, moving their equipment piecemeal up the mountain.

Those who made it over the summit crossed into Canada and continued on to Lake Bennett, the headwaters of the Yukon River. Later, a less-rough – but by no means easy – trail was discovered from Skagway through the White Pass to Lake Bennett.

At Lake Bennett, the stampeders had to build boats or rafts for the trip down the Yukon River to Whitehorse and Dawson City. Once the ice melted, the prospectors abandoned Bennett en masse and descended the river. Treacherous rapids claimed lives and supplies. At White-horse, riverboats would take them the rest of the way, while others continued on their own boats to the gold fields near Dawson City.

Canadian Mounties stationed at the border barred entry to the country to any party that did not have at least a ton of supplies, enough to last the whole year. Moving a ton of supplies over the trail was a monumental task. On the White Pass trail only a crude wagon road with several fords and a steep climb over the summit was available. It was along this arduous route that the White Pass & Yukon Railroad Co. sought to build a narrow-gauge line.

Surveying and construction

The 110-mile route began from the new tidewater port of Skagway, situated at the north end of Lynn Canal on the inland passage. The new line cut through coastal rain forests and past glaciers, clinging to ledges in vertical rock walls above the tree

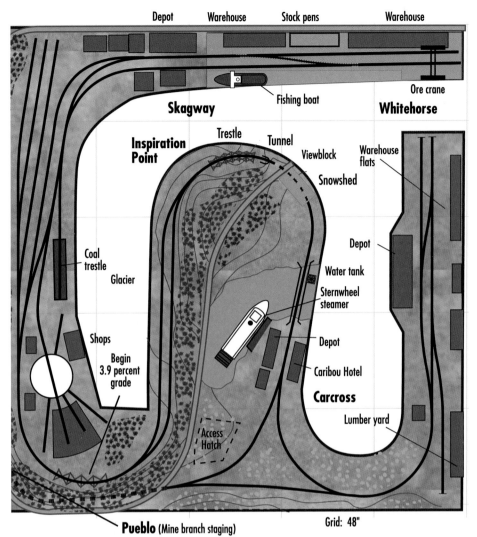

Depot   Warehouse   Stock pens   Warehouse

Fishing boat

Ore crane

Skagway

Whitehorse

Inspiration Point

Trestle   Tunnel

Viewblock

Warehouse flats

Snowshed

Coal trestle

Glacier

Depot

Water tank

Sternwheel steamer

Shops

Depot

Begin 3.9 percent grade

Caribou Hotel

Carcross

Lumber yard

Access Hatch

Pueblo (Mine branch staging)

Grid: 48"

LAYOUT AT A GLANCE

Overall size: 20 x 20 feet
Length of main line: 100 feet
Locale: Alaska
Layout style: walkaround
Layout height: 48"-53"
Benchwork: open grid
Roadbed: ¾" birch plywood
Track: code 125 handlaid
Turnouts: No. 5 handlaid
Minimum radius: 36"
Scenery construction: plywood surfaces,
    plaster over wire screen terrain
Backdrop: sheet styrene on wood frame
Control system: DCC with sound

White Pass & Yukon in G scale, circa 1942

---

line. At the summit near the U.S./Canadian border, the route descended through boggy taiga and permafrost tundra to Lake Bennett. It then followed the drainage of the Yukon River to the river port of Whitehorse. The route was quite steep, with an average grade of 2.6 percent and a maximum of 3.9 percent.

Work on the railroad commenced in April 1898. Amazingly, construction continued full scale through the winter, albeit behind schedule because of the harsh weather, solid granite bedrock, and a workforce distracted by the slightest rumor of new gold finds. By July 1899, the rails reached Lake Bennett. Unfortunately, by this time the gold rush was largely over. The good sites had been claimed, and the easily panned gold was exhausted. Nonetheless, the first train from Lake Bennett to Skagway carried $500,000 in gold dust.

Work continued from both ends of the line, and in June 1900 the railroad connected both Carcross and Whitehorse. In 1901, a dramatic steel cantilever bridge spanned the 215-foot-deep Dead Horse Gulch, replacing a steep switchback. At the time, it was the highest and northernmost bridge of its type. In the meantime, large companies were buying mineral rights and building mines with mechanized equipment to extract the remaining ore. With no work available, most prospectors left for home or other gold finds.

Over the next 40 years, the railroad struggled to remain profitable. Because there were no roads, the railroad was the main artery for freight traffic to the few miners working in the Yukon. Copper and other precious metal ores provided a modicum of steady traffic. The railroad built a pier at Skagway to help move the ore from car to ship.

Tourists were always a significant part of the WP&Y business. In the 1920s the railroad carried one tourist for every two tons of cargo. Unfortunately, the Great Depression caused the passenger business to dry up in the 1930s.

WWII revival

The railroad would have probably faded away if not for World War II. The threat of a Japanese invasion of Alaska made the WP&Y an important strategic asset. The railroad was overwhelmed by the wartime activity. The U.S. Army took over operation of the railroad via a lease arrangement, and its roster swelled to 36 locomotives and nearly 300 freight cars.

Early in the war, the U.S. and Canadian governments agreed to build the Alcan highway, which would connect Fairbanks to the mainland states over an inland route through Alaska and Canada. The

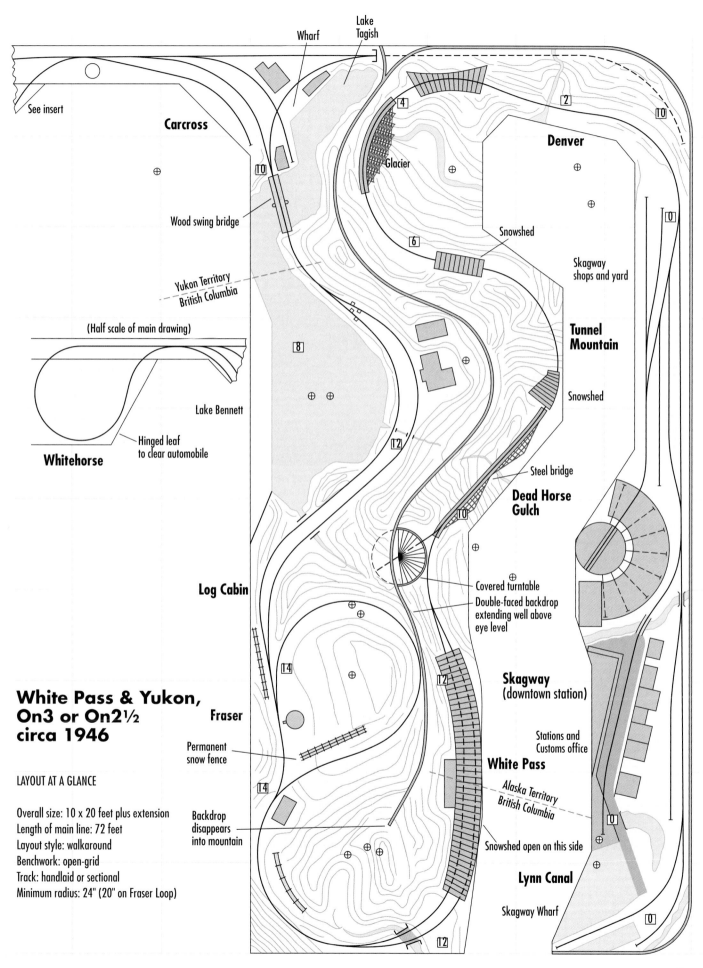

Wharf

Lake Tagish

See insert

Carcross

Wood swing bridge

4

Glacier

Yukon Territory
British Columbia

(Half scale of main drawing)

Lake Bennett

Hinged leaf
to clear automobile

Whitehorse

8

Denver

2

10

Snowshed

6

0

Skagway
shops and yard

**Tunnel
Mountain**

Snowshed

12

Steel bridge

10

**Dead Horse
Gulch**

Covered turntable

Log Cabin

Double-faced backdrop
extending well above
eye level

14

12

Skagway
(downtown station)

Fraser

Permanent
snow fence

Stations and
Customs office

White Pass

14

*Alaska Territory
British Columbia*

Backdrop
disappears
into mountain

Snowshed open on this side

Lynn Canal

Skagway Wharf

12

0

White Pass & Yukon,
On3 or On2½
circa 1946

LAYOUT AT A GLANCE

Overall size: 10 x 20 feet plus extension
Length of main line: 72 feet
Layout style: walkaround
Benchwork: open-grid
Track: handlaid or sectional
Minimum radius: 24" (20" on Fraser Loop)

Train No. 1 arrives at Whitehorse station in the 1950s, hauling several automobiles on flatcars. For many years the railroad was the only way to move autos from Whitehorse to Skagway. Number 81, an Alco 2-8-2, was built by Alco in 1920 as Sumpter Valley No. 19. The WP&Y bought the locomotive in 1942. *F.L. Jaques*

route was selected because it was out of range of Japanese bombers. The WP&Y proved vital in supplying workers and materials for the construction of the highway as well as for a pipeline from various oil wells to Whitehorse.

At its wartime peak, the railroad ran dozens of trains per day. In 1943, the WP&Y carried 280,000 tons of freight to Whitehorse plus thousands of troops and workers in both directions.

After the war, the development of natural resources in the Yukon – along with a rebirth in tourism – kept business at a steady level.

In 1954, the railroad began to dieselize. In 1955, the WP&Y continued its tradition of innovation by introducing ore containers and pioneering the construction of the first container ship, the *Clifford J. Rodgers*. With modern container ships and trucks, the WP&Y matured into a fully integrated transportation company of 1,000 employees.

In 1982, a worldwide drop in metal prices forced the mines in the Yukon to shut down, and the WP&Y – dependent

on the business – followed. The WP&Y didn't remain closed for long. In 1988 the railroad reopened exclusively as a passenger-carrying excursion line. It now hauls more than 200,000 tourists per year, mostly from cruise ships that include Skagway as a port of call.

Modeling the WP&Y

In March 1953, noted model railroad planner and author John Armstrong wrote in *Model Railroader* magazine, "The White Pass and Yukon happens to be unusually adaptable to modeling . . . its curves are too sharp, a 3.9 percent ruling grade keeps its trains short, and its surrounding scenery is space-saving vertical and changes with model-like abruptness in crossing timberline twice in a few miles. Yet it is a real railroad which proudly does a man-size job and sticks to big-time operating practices where it doesn't conflict with the peculiar problems of its traffic, terrain and climate." Armstrong's article included a ground-breaking track plan for an O scale version of the White Pass (see page 9).

Since that time, several ready-to-run models have become available based on (or usable for) the WP&Y. We'll explore modeling this amazing narrow gauge railroad in G scale.

The track plans

The Armstrong plan did a great job of distilling the key features of the railroad into a model layout. He may have even planted the seed for the Layout Design Element (LDE) concept when he stated, "don't overlook the advantages in realism that can come from careful study of a real railroad and its adaptation to your space." The plan's walkaround design and prototype basis show that John was well ahead of his time in designing track plans. His plan focused on the section from Skagway to Carcross because of the scenic opportunities. A staging loop represented the less-scenic but operationally more interesting portion to Whitehorse.

Armstrong's plan was designed for O scale and he suggested the plan could also be built in HO scale. However, he noted that because of the walkaround feature,

Snow, and lots of it, keeps WP&Y plows busy. Here the rotary plow clears the tracks at the loop at Fraser. *Karl Mulvihill*

Mikado No. 81 leads a freight downhill next to a masonry retaining wall around 1950. *F.L. Jaques*

Engine 196 and its train drop down the Skagway River near Reid's Falls. Number 196, a 2-8-2, was built for the U.S. Army in 1943. *F.L. Jaques*

the layout would not shrink in proportion, and if built to the same actual size in HO standard gauge, it "would become typical of a heavy-duty single-track line plagued with a tough climb." John even went so far as to design an HO track plan for a different freelanced prototype using the same overall shape. Recent layout planning concepts would probably not forgo the WP&Y in HO. Instead one would take advantage of the extra space to make an even more realistic rendition of the route.

Since the year 2000, manufacturers have released several new ready-to-run models of WP&Y equipment, especially in Sn3 and large scales. The Sn3 ready-to-run offerings focus on the post-1947 era. Converting Armstrong's plan to this scale would be relatively simple. You could build the layout in exactly the same footprint, adjusting only for track spacing and turnout size. The slightly smaller scale would allow more room for scenery and structures.

I chose to look at the other end of the spectrum and design a large-scale layout. Large-scale models, although often

referred to as G scale, are not fully standardized. The one common aspect is that all use 45mm-gauge track. Manufacturers offer models in various proportions including 1:32, which represents standard gauge (No. 1 scale), 1:20.3 for three-foot narrow gauge (G or Gn3), and 1:22.5, which is proper for European meter gauge but often used for American narrow-gauge prototypes as well. To further confuse the issue, some models are made in 1:24 or 1:29 as well. Stick to narrow-gauge prototypes and models in 1:20.3 or 1:22.5 to keep proportions consistent.

An indoor large-scale layout?

Although large-scale layouts are commonly associated with outdoor garden railroads, the advantages of building an indoor fine-scale model railroad that is operationally oriented are worth considering. With an indoor layout, structures and scenery can be made without regard to effects of adverse weather. Those interested in fine-scale modeling can build exquisitely detailed models without worrying about attacking cats, termite invasion, vandalism, or heavy rainstorms.

Without the effects of weather, one can devote considerable attention to modeling the track closer to scale standards. Most 45mm sectional track is oversized, using rail of code 250 (.250" tall) or larger so it will be durable outdoors. This is too hefty for accurately modeling the narrow-gauge rail used on the real WP&Y. Handlaying code 125 or 148 rail, which is normally associated with S and O scales, will result in track that is much more realistic than normal sectional track.

Building the railroad on standard benchwork at a convenient height instead of on the ground offers a better opportunity to appreciate the fine detail and smooth-running locomotives available in G scale. Watching near eye level as side rods flash and listening up close as the sound system chuffs will be a real treat, one that garden railroaders largely miss.

Because of their mass, large-scale engines tend to run very well, especially at slow speed. Even diminutive narrow-gauge engines are large in G Scale. The size allows for big speakers, so they have wonderful sound systems and great steam locomotive animation. Furthermore,

This photo was taken at Carcross in 2005, but other than a road replacing the rails, the scenery remains largely the same as when the WP&Y was built.

keeping the track clean will be much easier indoors. Another advantage is that G scale cars can be reliably backed, and operation is quite smooth.

Designing an indoor layout for G scale is not without its challenges. Everything is big – a No. 5 turnout is about 22" long! Thus, any attempt to model a prototype line will require selective compression. Although the schematic of the layout is compressed, the actual distance is quite long, so the layout will not feel small.

Strict prototype-based modeling can be difficult in G scale due to the lack of many fine-scale items and the varying scales used by manufacturers. Some compromise in era and equipment will be necessary. There is a good, but not complete, selection of WP&Y equipment available, so modelers attempting to model this road need not scratchbuild everything. However, scratchbuilding finely detailed models is a large part of what this layout is about.

Layout features

The plan represents in highly compressed form the 110-mile line from Skagway to Whitehorse. Some key features had to be omitted, such as the steel bridge over Dead Horse Gulch, but the major operational aspects are represented.

While selectively compressed, the layout is still large. Though the mainline run is only about a half of a scale mile, it is 100 actual feet, not much less than a medium-to-large-size layout in a smaller scale and a good distance for a walk-around layout. Running a train from Skagway to Whitehorse should give the impression of a long distance.

The layout maintains a 48" minimum radius except for the turn-back loop, which is 36", but much of it is hidden in the tunnel. If more room is available, a larger radius here would be beneficial for appearance, especially on the trestle. The plan uses No. 5 turnouts. As mentioned earlier, these will have to be handlaid, including the curved turnouts at Inspiration Point. However, only a few turnouts are needed, so the task should not be overwhelming.

Like any railroad, the track arrangements for the WP&Y changed over the years. The wharf with its tracks was

rebuilt several times. The tracks through town were relocated to the east along the mountain slopes after World War II. The dramatic steel bridge was taken out of service in 1969, and a tunnel was added in the new alignment. Turntables and wyes came and went. The layout plans combine features from the various eras to maximize operational potential. I would set the railroad in the WWII era, modeling it during its busiest period.

The track plan of the wharf at Skagway represents the WWII-to-1970s configuration when the tracks were adjacent to the waterline. There's no room to actually model the steep mountain behind the warehouses, so it will have to be painted on the backdrop. A model of a tramp steamer would probably be too large to fit at the pier, but there's room for a smaller fishing vessel. During wartime, construction cargo cluttered the piers.

The plan includes a brief section of Skagway when the tracks ran right down Broadway in the middle of town. The station has been restored and is still used by the tourist railroad as well as a museum. Building this as well as other quaint

Stevedores unload a UTLX tank car and prepare to place it on the rails at the wharf in Skagway. The mountain rises sharply behind the wharf. *Carl Mulvihill*

structures connected by a wooden sidewalk will provide hours of modeling fun.

The engine terminal is very compressed. The roundhouse has only two stalls versus the 25 of the prototype. The coal dock and other shop buildings fill in the remaining space. The yard is designed for a limited amount of switching and train building.

Inspiration Point on the layout is a composite of several locations on the actual line. It is hard to imagine a more spectacular scene than floor-to-ceiling scenery where the trestle approaches the tunnel. The tunnel and snow shed hide where the tunnel pierces the skyboard.

A central divider skyboard divides the layout into two sections. The southern part depicts coastal rain and sub-alpine forests. Tundra, bogs, lakes and rolling hills are appropriate for the Carcross-to-Whitehorse section. Appropriate scenes painted on the backdrop will extend the landscape.

The small town of Carcross has some irresistible features. The swing bridge with water tank, Caribou Hotel, and depot are still standing. The plan includes

all of these. A model of a paddle-wheel steamer at the dock would be a fun project. Alternatively, a good artist could paint the steamer on the backdrop as if it were approaching from the distance.

This plan omits a separate junction at Macrae, where a mine branch was built to Pueblo. However, the tail track at the Carcross wye extends under the hills on the other side of the backdrop. This track will do double duty since it is long enough to use as a staging track for Pueblo as well as a means to turn engines.

As on the prototype, Whitehorse is the end of the line. A removable track across the aisle connecting Skagway to Whitehorse for continuous run is possible but not needed for realistic operations. The wye between Carcross and Whitehorse can be used to turn locomotives for the return trip. There are several industries at Whitehorse. The town was also a river port, and the paddle-wheel steamers should be painted on the backdrop.

Layout construction

You can use conventional indoor layout construction techniques for this layout.

Make sure the benchwork is sturdy, as you may have to climb on the wider sections during layout construction and later for maintenance. Some areas such as Whitehorse and Skagway can be built on flat pieces of ¾" birch plywood with risers every 12" to 16". In the mountain sections, spline roadbed made with 1" pine lath on edge might be more appropriate.

Most of the turnouts on the WP&Y were stub switches, which are not available in G scale. However, supplies for handlaying are available. Micro Engineering offers code 125 and code 148 rail and spikes. Several manufacturers offer scale lumber ties and cast switch stands for actuating turnouts.

Many large scale cars and locomotives have wheels with deep flanges. These may need to be replaced if you're using smaller rail. NorthWest Short Line and Paragon-Aero offer realistic-face-profile wheelsets with scale and semi-scale flanges.

The large mountains behind Inspiration Point will require special attention. Plaster on screen wire would work. However, carving expanded foam might be faster and will be lighter in weight.

Glacier Gorge and Tunnel Mountain loom high above the tracks on the south side of the valley.

Building large-scale dioramas has taught me that scenery materials different from those in smaller scales are needed. Sifted soil, cat litter, and small bits of rock make realistic basic terrain texture. For ground cover, static grass, fake fur, and selected natural weeds work well. Ground foam doesn't work well in large scales.

Individual trees can be modeled using carved wood trunks and natural weeds for limbs. Photo-etched leaves are available to represent specific species. Making each tree is time consuming, but each covers a lot of ground, so fewer are needed. As the tracks climb in elevation, the surrounding trees are stunted, making modeling easier.

Locomotives and rolling stock

During the war, the WP&Y had an all-steam roster, running a wide variety of 2-6-0, 4-6-0, 2-8-0, and 2-8-2 locomotives. Many were acquired second-hand from other railroads such as the East Tennessee & Western North Carolina and Denver & Rio Grande Western. Bachmann offers 2-8-0 and 4-6-0 engines that are quite acceptable out of the box, and they could easily be modified

to look more like WP&Y prototypes. LGB once offered an excellent limited-edition WP&Y 2-8-2. The real WP&Y didn't receive these until 1947, but they are still in use today.

There is a limited selection of ready-to-run WP&Y freight and passenger cars. However, many appropriate cars are available if one doesn't mind repainting.

Operations

The layout should be run using time-table-and-train-order operation. There are three sidings and the wye at Carcross where meets are possible. The 1949 prototype schedule shows only one mixed train in each direction two times a week, usually timed to meet incoming steamers. Other trains ran as extras. However, during World War II the railroad was much busier, with more than 30 engines used on dozens of trains per day.

This layout could easily support four or five operators. One could work the yard and hostle engines, and the others can run freights and mixed trains on the main line. Car types include passenger, ore, lumber, boxcars, and flats carrying

jeeps, trucks, autos, and boats. The design train is six feet long – one engine and four or five cars – to match the length of the siding at Whitehorse. The sidings at Skagway and Inspiration Point are longer to allow using a helper engine. Creative switching will be required to perform a meet at Whitehorse if a longer train is run all the way there. Using a curved turnout near the north wye turnout could lengthen the siding at Whitehorse.

Note that this layout has no rail inter-changes or staging tracks except the branch at Pueblo. There were no connecting rail lines; only steamships awaited the trains at both ends.

Since ore in gondolas was an important cargo, removable model loads will be needed for cars carrying ore to Skagway. However, ore was shipped to and from steamships at Whitehorse. Operations can be planned accordingly.

This layout packs lots of realistic operation into a relatively simple track plan. Combine that with the thrill of operating large-scale trains over spectacular terrain, and you have a combination that is hard to beat.

CHAPTER 2

Western Maryland's Thomas Subdivision

The station and neighboring bridge at Thomas, W.Va., make up a signature scene on the Western Maryland's Thomas Subdivision. Here Alco RS-3 No. 187 and EMD F7 No. 64 lead their train through the scene. *Tony Koester*

Sheer cliffs, tumbling waterfalls, and rock-laced rapids caused passengers on the Western Maryland's Thomas Subdivision to marvel as their trains traveled through the Appalachians from Cumberland, Md., to Elkins, W.Va. The line's steep grades, sweeping curves, interesting but manageable operations under timetable and train order control, and readily available engines and rolling stock in N and HO scales make the railroad an attractive subject for a medium-size layout.

The Western Maryland's Thomas Subdivision

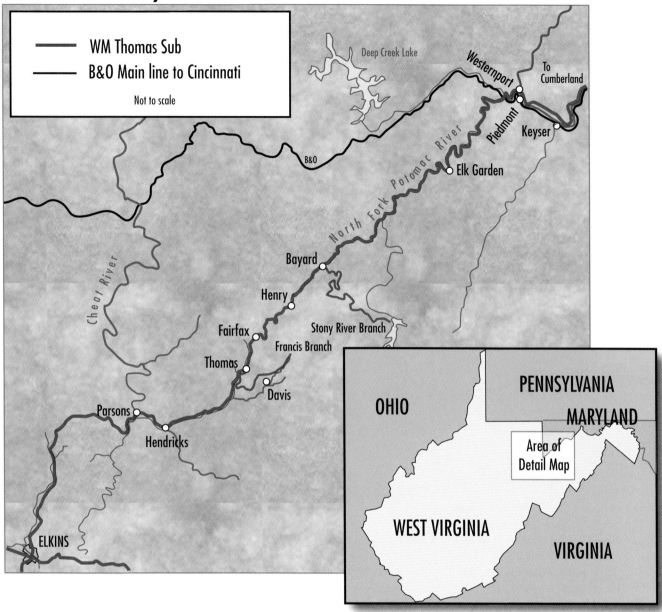

History

The Thomas Subdivision got its start as the Potomac & Piedmont Coal and Railroad Company, which began building a line along the North Branch of the Potomac River in 1880. For the next four years the company hauled coal from the Potomac Highlands around Davis and Thomas to its interchange with the Baltimore & Ohio at West Virginia Central Junction in Piedmont, W.Va.

The B&O did not appreciate the competition and attempted to stifle this traffic. Instead, the Potomac & Piedmont, now called the West Virginia Central & Pittsburg, built its own line between Piedmont and Cumberland. In 1888, the railway expanded southward, following the watershed of the Potomac River as far as Henry, W.Va. Upon climbing to Fairfax, W.Va., it entered the Cheat River Watershed near Thomas, W.Va. Continuing south, the line dropped down the steep and scenic Blackwater Canyon to Parsons. One more ridge, Cheat Mountain, had to be climbed until the line reached the resource-rich area near the town of Leadville, later called Elkins, in 1889.

With steady coal, timber, and tanning bark traffic, the financial success of the railroad generated a steady stream of revenue, a tempting target for the rail barons of that age. In 1902, George Gould took control of the WVC&P by ruthlessly buying out the partners. Gould's dream was to construct a coast-to-coast railroad. To do this he needed cash, and the revenue from the profitable WVC&P could be a source. However, after four turbulent years, Gould's financial troubles crushed his transcontinental ambition.

Another of Gould's lines, the bankrupt Western Maryland Railroad, absorbed the WVC&P in 1906, and in 1910 the refinanced Western Maryland Railway took over. For the next 60 years, the WM operated the profitable line, hauling coal, wood products, and tanning goods out of the hollows of West Virginia. The WM rebuilt the line and upgraded it to secondary mainline status, which it maintained well into the early 1970s.

The single-track subdivision ran 112 miles between the imposing brick station in Cumberland and the more modest brick station in Elkins. Timetables governed operations, with dispatchers issuing train orders that were passed along at train order offices along the line. Dispatchers were kept busy controlling eastbound trains and the multiple helper sets needed to surmount the 3.05 percent grade in Blackwater Canyon.

The lines stayed in operation through the Chessie System mergers, but in 1978 Chessie began to eliminate redundant routes, and much of the Thomas Sub was closed. Today under CSX only the section from Cumberland to Henry is still active, serving the few coal loaders and pulpwood sources still in operation. The scenic route through Blackwater Canyon is now a bicycle trail.

Equipment
During the steam era, the Thomas Sub was almost exclusively the home of 2-8-0 Consolidations, while 4-6-2 Pacifics handled the passenger trains. Diesel switchers could be found working the yard at Elkins as early as 1947, but it was the arrival of the WM's initial first-generation diesel road locomotives, Alco RS-2s and RS-3s, in the early 1950s that quickly sent the steam locomotives into retirement. The few years when steam and diesel served the line at the same time is the time period depicted in these layout plans.

In later years, EMD F7s and first-generation Geeps (GP7s and GP9s) also handled trains on the subdivision. The sharp curves found throughout the main line and on the other subdivisions south of Elkins ruled out larger motive power.

Thomas Sub characteristics
The WM maintained a medium-size yard, an extensive car repair shop, and a brick station at Elkins. From there, other subdivisions radiated south to coal fields and lumber mills. Also at Elkins, the B&O and WM interchanged freight from the west. The engine terminal featured a large roundhouse and a Roberts and Schaefer wooden coal dock.

From Elkins, eastbound trains had to tackle the steep grade up Blackwater

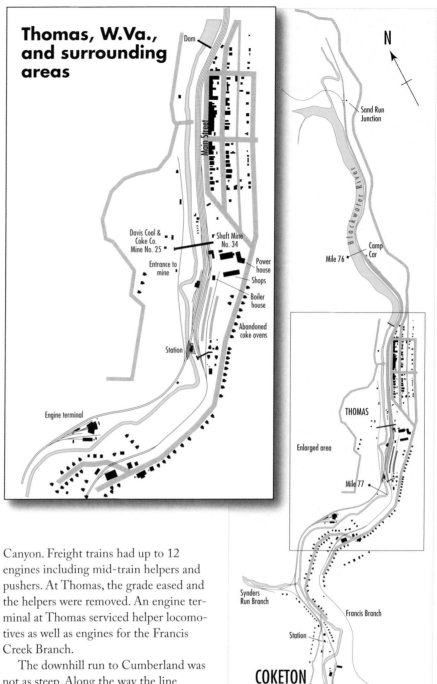

Thomas, W.Va., and surrounding areas

Canyon. Freight trains had up to 12 engines including mid-train helpers and pushers. At Thomas, the grade eased and the helpers were removed. An engine terminal at Thomas serviced helper locomotives as well as engines for the Francis Creek Branch.

The downhill run to Cumberland was not as steep. Along the way the line served several coal mines as it followed the Potomac. At Westernport the West Virginia Pulp and Paper Company maintained a large mill, with both the B&O and WM providing rail service. After passing these facilities, the Thomas subdivision ended at Ridgely, W.Va., near Cumberland.

The summit at Thomas has much to offer both scenically and operationally. The unique station built astride the Blackwater River, several coal mines, and a classic main street facing the main line are all attractive elements to model.

N scale plan
The N scale design is about 8 x 11 feet, yet its roadbed can be fashioned from a single 4 x 8-foot sheet of plywood. It uses a single-track oval plan that emphasizes through coal train and helper action. The primary pattern for coal traffic is loads east, empties west. An oval plan with a hidden portion is well suited for this, since trains that end their run are imme-

The yard at Thomas is visible through the fireman's windshield of a Western Maryland F7. The town is on the hill above the railroad. *Tony Koester*

The two lead diesels on this 72-car coal drag will cut off here at Thomas, and the two trailing 2-8-0s will take the train on to Cumberland. Farther back are five diesel helpers. The lead locomotive, No. 186, is an Alco RS-3 delivered in May 1953, and the second, No. 192, is an RS-2 built in April 1950. *William N. Poellot Jr.*

Consolidation No. 837 leads this mixed freight train, with several pushers working at the rear. The class H-9 2-8-0 was built by Baldwin in 1921. *George C. Corey*

diately restaged for the next run. Manifest freights and locals, by virtue of their enclosed cars, can run in either direction.

Virtually all the visible tracks are within yard limits at Thomas. Three staging tracks serve this yard via small sections of main line. The staging tracks are

hidden, but accessible from the rear (see cross section AA for staging access). Sidings and staging tracks are long enough to handle trains of three locomotives and 20 to 25 50-ton hopper cars.

With just over a half scale mile (18 actual feet) of visible track, the plan can't

capture the full nature of the prototype's 112-mile subdivision. However, by focusing on the signature scene at the top of the grade at Thomas, it still captures the drama and excitement of this mountain railroad.

Western Maryland crews removed helpers at Thomas, and this operation is a

Thomas Subdivision, N scale basic plan, circa 1952-53

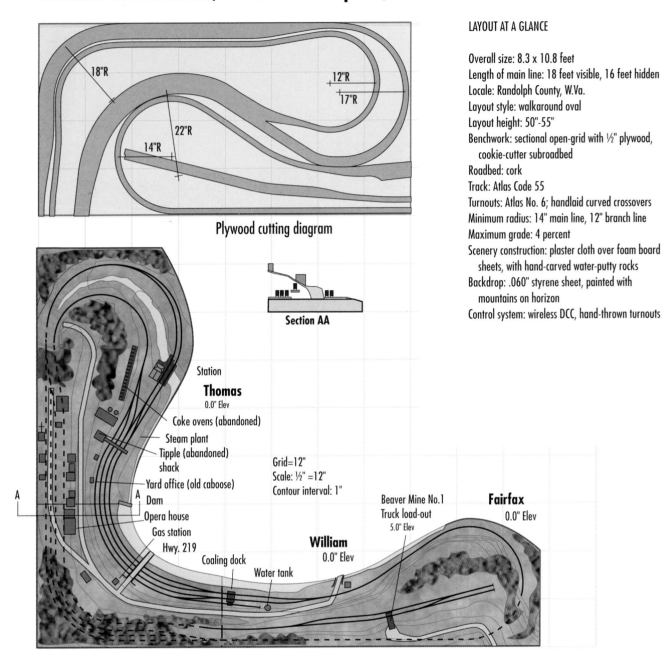

Plywood cutting diagram

18"R

12"R

17"R

22"R

14"R

Section AA

Station

Thomas
0.0" Elev

Coke ovens (abandoned)

Steam plant

Tipple (abandoned)
shack

Yard office (old caboose)

Dam

Opera house

Gas station

Hwy. 219

Coaling dock

Water tank

Grid=12"
Scale: ½"=12"
Contour interval: 1"

Beaver Mine No.1
Truck load-out
5.0" Elev

Fairfax
0.0" Elev

William
0.0" Elev

A

A

LAYOUT AT A GLANCE

Overall size: 8.3 x 10.8 feet
Length of main line: 18 feet visible, 16 feet hidden
Locale: Randolph County, W.Va.
Layout style: walkaround oval
Layout height: 50"-55"
Benchwork: sectional open-grid with ½" plywood,
      cookie-cutter subroadbed
Roadbed: cork
Track: Atlas Code 55
Turnouts: Atlas No. 6; handlaid curved crossovers
Minimum radius: 14" main line, 12" branch line
Maximum grade: 4 percent
Scenery construction: plaster cloth over foam board
      sheets, with hand-carved water-putty rocks
Backdrop: .060" styrene sheet, painted with
      mountains on horizon
Control system: wireless DCC, hand-thrown turnouts

key feature of the layout. Most eastbound freight trains had mid-train and rear helpers. Here these would be cut off, the train reassembled and sent east, and the helpers returned west. Duplicating this on the layout can be achieved by providing access to the hidden staging yard during operating sessions. There a "fiddler" operator can set up trains with appropriate helpers. Operators on the other side would take over to execute the maneuvers described above.

The coal dock and water tower are correctly placed to allow realistic opera-

tion when refueling and watering steam engines.

The Francis Branch joined the main line at Thomas. Including it on the layout provides a chance to operate a mine shifter based out of Thomas to work the truck load-out at Beaver.

A layout's long-term satisfaction is enhanced if it is designed for easy expansion. This layout is not designed to modular standards, but by following modular techniques at the section joints it would be easy to expand this plan in a prototypical manner. The drawings show two pos-

sibilities depending on available space. Expansion would allow an even more realistic depiction of Thomas.

Once built, these three sections could form the core of a larger layout that could include more of the subdivision.

The layout has space for three trains in staging and one on the branch. A typical operating session would have a dispatcher control the staging while one or two operators run trains across the layout. Since most of the layout is in yard limits, the dispatcher will control the yard track assignments and call crews. Nearly all

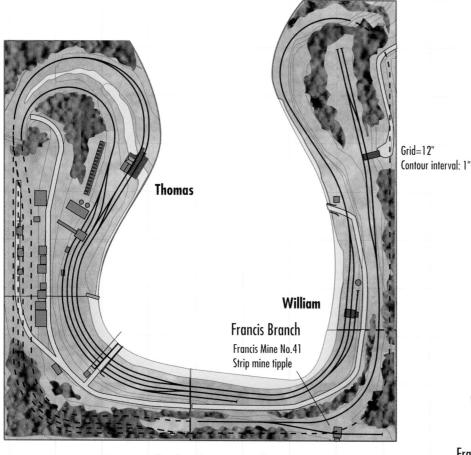

Thomas

William

Francis Branch

Francis Mine No.41
Strip mine tipple

Grid=12"
Contour interval: 1"

Expanded plan, U shape

N scale

William

Francis Branch
5.0" Elev
Francis Mine No.41
Strip mine tipple

Expanded plan, L shape

Thomas

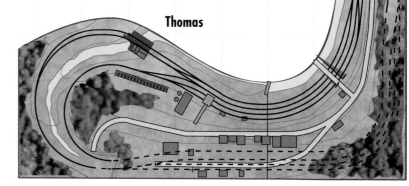

The expanded N scale plans share many of the same features as the basic plan, but with an expanded main line and additional switching and helper operations.

prototype trains used helpers, which will provide a challenge to operating crews.

The short mainline run will seem longer if crews run their trains at prototypical speeds. Coal drags climbed the summit at 12 mph, meaning model trains will take several minutes to cross the layout.

Thanks to the oval design, trains finish their run by returning to staging, although some fiddling will be necessary to restage the helpers. Thus each train can run at least twice during a session. The dispatcher/yardmaster can also fiddle cars between runs, providing variety to the consists. The yardmaster's job would be made easier with shelves or drawers below the yard to hold trains between runs.

The HO plan

Designed for a 10 x 20-foot space, the HO plan includes all the features of the N scale plan, plus a long stretch of the scenic Blackwater River Canyon.

The staging area is located under Thomas. It's shown unscenicked in the drawing, but scenery could be added to simulate Elkins.

This layout could easily support three crew members: a through train operator, a helper, and a local switcher at Thomas. The helper crew ties on to the train at Elkins. The head-end and helper crews

Thomas Subdivision, HO Scale, circa 1952-53

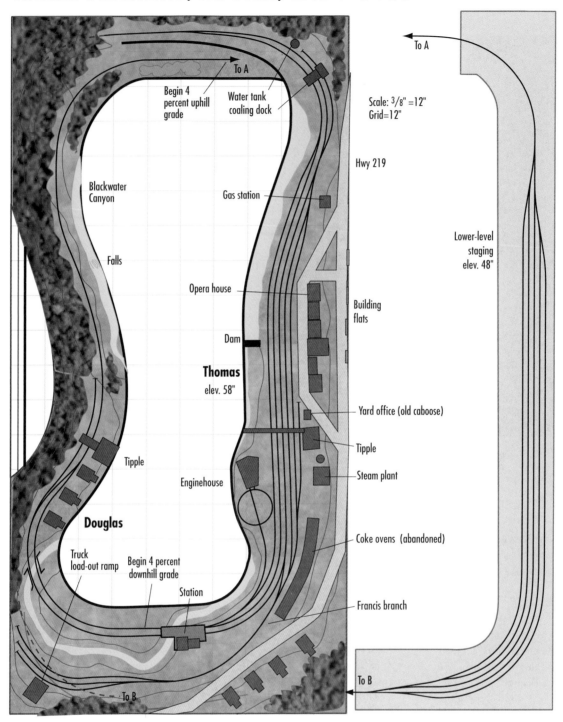

To A

Begin 4 percent uphill grade

Water tank coaling dock

Scale: $^{3}/_{8}$" =12"
Grid=12"

Hwy 219

Blackwater Canyon

Gas station

Falls

Opera house

Dam

Thomas
elev. 58"

Building flats

Yard office (old caboose)

Tipple

Steam plant

Tipple

Enginehouse

Douglas

Truck load-out ramp

Begin 4 percent downhill grade

Station

Coke ovens (abandoned)

Francis branch

To B

To A

Lower-level staging elev. 48"

To B

LAYOUT AT A GLANCE

Overall size: 10 x 20 feet
Length of main line: 50 feet visible, 20 feet hidden, 24 feet staging
Locale: Randolph County, W.Va.
Layout style: walkaround oval

Layout height: 48"-58"
Benchwork: sectional open-grid
Roadbed: cork on ½" plywood, cookie-cutter style
Track: Atlas code 83
Turnouts: Atlas No. 6, handlaid curved turnouts

Minimum radius: 30" main line, 24" turntable lead
Maximum grade: 4 percent
Backdrop: .060" styrene sheet
Control System: Wireless DCC, hand-thrown turnouts

take the train up the steep 4 percent grade through the Blackwater River canyon. The train arrives in Thomas and the helpers cut off.

The through train continues to the coal dock, where it takes water and coal, then returns to staging via the hidden return track. Once there, it is ready for another run. A typical session would include one or two coal trains, a mixed freight, and a passenger train.

In the meantime, a local switch crew would assemble its train in the Thomas yard, also working the coal mines and the house track at the coal dock. The mine in the real town was abandoned in the 1950s, but assuming it is still active adds more operating interest. The truck load-out on the Francis Branch requires some interesting car sorting to pull the loads and spot empty hoppers.

The plan includes a turntable, even though the prototype didn't have one. Real engineers used the wye at Snyder's Run to turn helper engines for the return to Elkins, but the turntable saves space on the layout.

One steam locomotive was needed for every 10 loads on the steep grade up Blackwater Canyon. The two lead and three helper engines display their effort in this 1951 view. *William P. Price*

Two Consolidations, led by No. 821, fight upgrade at the head end of a coal train near William, W.Va., in July 1953. *William N. Poellet Jr.*

CHAPTER 3

Norfolk Southern's Shirley Industrial Park

Two venerable Norfolk Southern GP38-2s shove a cut of boxcars into a warehouse spur at the Shirley Industrial Park in Springfield, Va., in 2006.

Switching districts are good prototype subjects for mid-sized layouts, especially one designed to fit a spare bedroom. And, if the prototype includes a variety of industries serviced by 60-car trains powered by the latest motive power, all the better. Such is the case for the Shirley Industrial Park, a switching complex served by the Norfolk Southern in Springfield, Va.

Norfolk Southern in Northern Virginia and Washington, D.C.

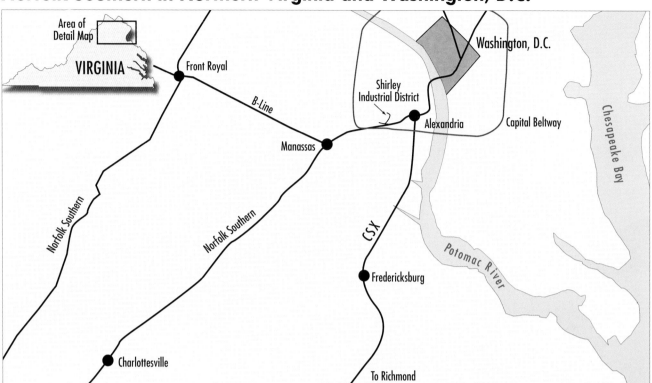

More than hot air and paper

Prior to World War II, most of northern Virginia's Fairfax County was rural and had a distinctly southern flavor, with dirt roads, small farms, and very little industry. After the war, the area experienced a boom and quickly developed as a bedroom suburb of Washington, D.C. It is not an area known for heavy industry, as the nation's capital's primary output is hot air and paper. Nonetheless, several businesses, primarily construction material suppliers, light manufacturers, and warehouses, developed along what is now the Norfolk Southern from Alexandria to Manassas.

The NS in northern Virginia traces its roots to the Orange & Alexandria Railroad. Chartered in 1848, the O&A was the first railroad built on the south side of the Potomac River near Washington. Construction of the railroad began at its headquarters in Alexandria, proceeded through Tudor Hall (later renamed Manassas Junction) and Charlottesville, finally reaching Lynchburg by 1860.

These lines played a critical role for both sides in the Civil War. The northern portion from Alexandria to Manassas was under the control of the United States Military Railroad and helped supply the

At the Vulcan Materials facility, two of the company's unusual two-bay rapid-discharge hopper cars sit under a conveyor in 2001. Large storage piles of aggregates can be found throughout the property.

Union army. The Confederates used the southern portions to supply Richmond and strategically deploy troops.

After the war, the Baltimore & Ohio and the Pennsylvania Railroad fought for control of the northern end of the O&A. This continued until the turn of the century, when Potomac Yard was established as a joint operation for six railroads with interests in reaching Washington.

Through a complicated series of mergers, the O&A became part of the

Southern Railway in 1894. The Southern ran the line for 88 years until merging with the Norfolk & Western in 1982 to form Norfolk Southern Corporation.

The Alexandria line remains under NS control today as part of the Washington District. The north end of the district is now at an interlocking with CSX named Val (formerly AF Tower) just south of the Alexandria passenger station. The Washington District continues south to Lynchburg and also includes the B-Line,

Shirley Industrial Park, Springfield, Va.

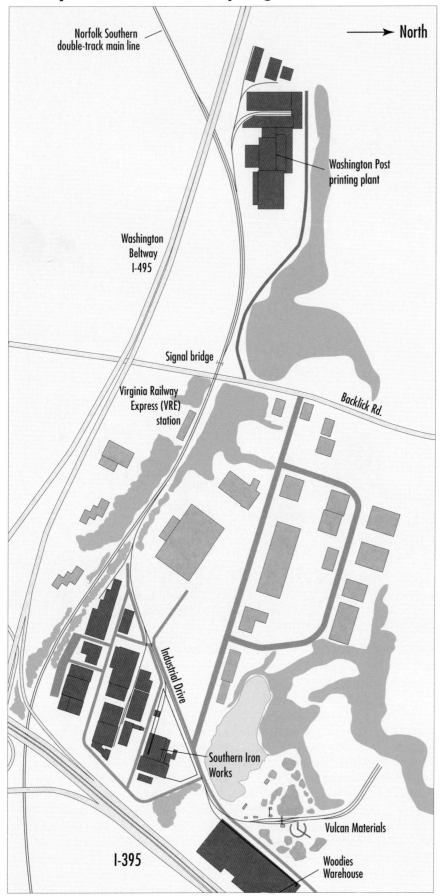

Norfolk Southern double-track main line

→ North

Washington Post printing plant

Washington Beltway I-495

Signal bridge

Virginia Railway Express (VRE) station

Backlick Rd.

Industrial Drive

Southern Iron Works

Vulcan Materials

I-395

Woodies Warehouse

which runs from Manassas to the Hagerstown District in Front Royal.

The main line is controlled by a dispatcher using CTC (Centralized Traffic Control) signals, while the B-Line uses dispatcher-issued track warrants. Most north- and southbound traffic on the Washington District uses the B-Line to avoid Alexandria and downtown Washington.

Between Alexandria and Manassas, where NS maintains a small yard, there are just a few scheduled local freight trains. These include a coal local that works from Manassas to the power plant and Robinson Terminal in Alexandria and back. This train can also switch the Shirley Industrial District warehouse. Other locals work from Vulcan Materials quarries in Virginia to two Vulcan facilities in Alexandria, one in Shirley, and the other near Van Dorn Street, across from the former intermodal yard. Intermodal trains once originated in Alexandria's Van Dorn Street Yard bound for Atlanta, and their northbound counterparts terminated here.

The line still sees passenger traffic. Virginia Railway Express commuter trains run on NS tracks from Alexandria to Manassas, and Amtrak's *Crescent* and *Cardinal* travel on portions of the Washington District departing Washington.

1950s industrial development

The Shirley Industrial District developed in the 1950s to provide a location for light and medium industries. The district is located in Springfield at the intersection of I-395, the Shirley Highway, and the then-under-construction Capital Beltway. The Southern main line was adjacent to the district's southern end and provided rail service to the park.

One of the first companies to move into the industrial park in the early 1950s was Southern Iron Works, a company specializing in structural steel fabrication that had been in the Washington area since 1933. The facility had a stub siding where gondolas brought raw steel plates into a receiving yard. Overhead cranes unloaded the steel plates and stored them in orderly piles at the yard until they were needed. The cranes then transferred the plates into the main structure where the fabrication is done.

Shirley Industrial Park, HO scale, circa 1980s

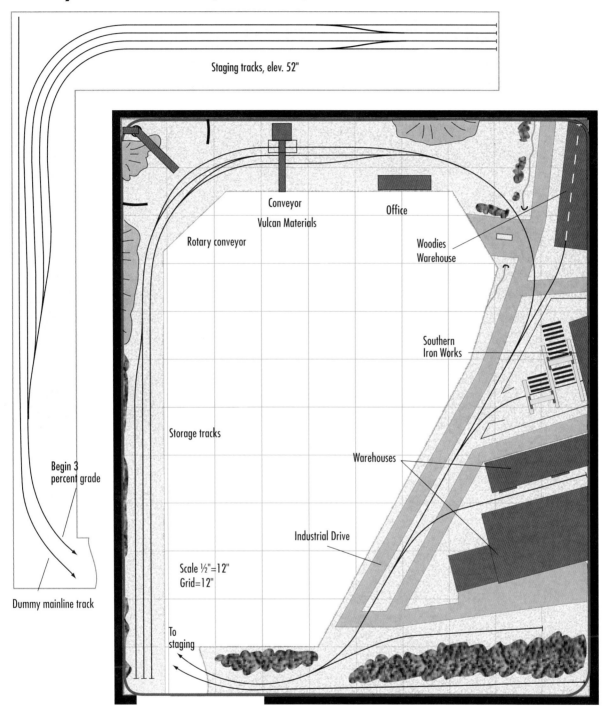

Staging tracks, elev. 52"

Conveyor

Vulcan Materials

Office

Rotary conveyor

Woodies Warehouse

Southern Iron Works

Begin 3 percent grade

Storage tracks

Warehouses

Dummy mainline track

Industrial Drive

Scale ½"=12"
Grid=12"

To staging

LAYOUT AT A GLANCE

Overall size: 10 x 12 feet
Length of line: 29 feet (visible)
Locale: Springfield, Va.
Layout style: walkaround, point-to-point
Layout height: 57"
Benchwork: sectional open-grid

Track: code 83
Turnouts: Atlas No. 6, handlaid curved turnouts
Minimum radius: 30"
Maximum grade: 3 percent
Backdrop: .060" styrene sheet
Control system: wireless DCC, hand-thrown turnouts

The rail siding at Southern Iron Works remains in place, but its rusty rails haven't seen a train since the early 1990s. The yellow beams support overhead cranes, and a semi is parked on the end of the spur track. A chain-link fence surrounds the facility. The outdoor steel storage yard would be an interesting model subject.

Vulcan has its own switcher (at left) to shuffle cars once the NS train has dropped them off. In the background are several of Vulcan's red Ortner hopper cars.

The finished steel beams are shipped to construction sites all around the Washington and Baltimore areas, usually by truck. The company stopped receiving steel via rail in the early 1990s, but the tracks and unloading facilities are still in place. For operating interest, the layout plans assume rail service has continued.

Robinson Terminal Warehouse has one of its five warehouse locations in the industrial park. The company has been in the warehousing business since 1939 in Alexandria, though it can trace lineage to an Alexandria firm as far back as 1853. The company handles many different commodities, but concentrates in forest products, especially newsprint for the *Washington Post*. The company's nearby terminal in Alexandria is one of the largest handlers of newsprint on the East Coast and has the only working port in the greater Washington metropolitan area. Ships carrying newsprint frequent the harbor.

The NS spots several boxcars at the Shirley Industrial Park warehouses each

Alternate HO plan with continuous loop

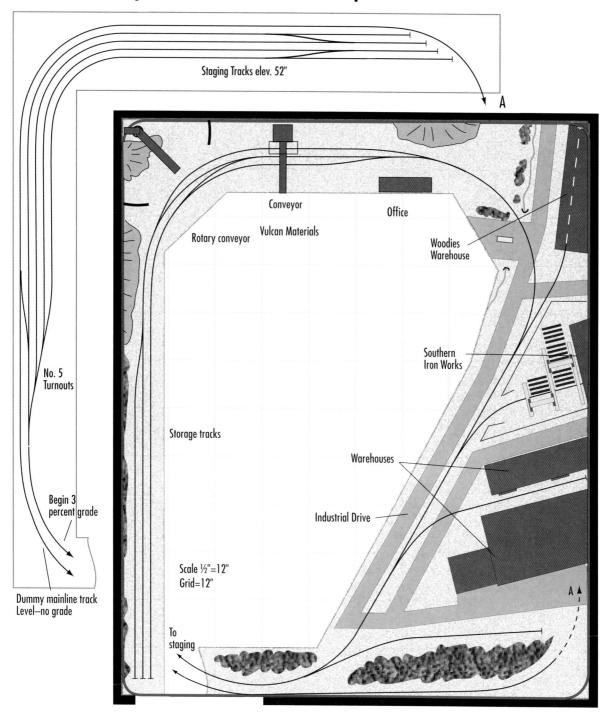

Staging Tracks elev. 52"

A

Conveyor

Office

Vulcan Materials

Rotary conveyor

Woodies
Warehouse

Southern
Iron Works

No. 5
Turnouts

Storage tracks

Warehouses

Begin 3
percent grade

Industrial Drive

Scale ½"=12"
Grid=12"

A

Dummy mainline track
Level—no grade

To
staging

LAYOUT AT A GLANCE

Overall size: 10 x 12 feet
Length of line: 29 feet (visible)
Locale: Springfield, Va.
Layout style: walkaround, point-to-point
Layout height: 57"
Benchwork: sectional open-grid

Track: code 83
Turnouts: Atlas No. 6, handlaid curved turnouts
Minimum radius: 30"
Maximum grade: 3 percent
Backdrop: .060" styrene sheet
Control system: wireless DCC, hand-thrown turnouts

The NS uses modern six-axle locomotives, such as these GE Dash-8s, when switching Vulcan Materials. The former Woodies warehouse is behind the locomotive at right.

week. Other cars are shuttled from the Alexandria wharf warehouse to the *Washington Post* plant.

The *Washington Post's* $60 million satellite printing plant in Springfield was built in 1983 to accommodate the newspaper's growing circulation and conversion to cold type. The plant was extensively remodeled and modernized in 1997 to accommodate new offset presses.

Designers and editors make up pages at the paper's downtown office, and when the pages are finished, they are electronically transmitted to the printing plant via a fiber optic connection. During a typical night, the plant's eight presses print about 900,000 newspapers in a four- to five-hour period. Trucks carry the papers to 450 distributors.

The plant receives five to eight boxcars of paper a day, seven days a week. The plant has four stub-end sidings where these cars can be spotted. They are usually switched at night. The paper comes from Vermont and Canada as well as the southern United States.

Vulcan Materials Company's Edsall Road Yard and Recycle is the largest rail customer in the Shirley park. Vulcan, headquartered in Birmingham, Ala., is the nation's leading producer of construction aggregates, primarily crushed stone, sand, and gravel, materials used in nearly all forms of construction. Across the country, Vulcan owns 166 stone quarries, 37 sand and gravel plants, 66 sales yards, 45 asphalt plants, and 22 ready-mix concrete facilities that shipped 255 million tons of aggregates in 2006.

Vulcan's facility at Shirley is a recycled concrete and gravel supply yard. The primary products are asphalt aggregate, base material, concrete aggregate, manufactured sand, and re-crushed concrete. The company has a three-track yard and operates its own small switcher. Several conveyors, crushers, and numerous material piles scattered about the facility will provide modeling challenges.

Norfolk Southern delivers one or two trains a day to Vulcan, each comprising up to 60 rapid-discharge hopper cars. The

cars, many of which bear Vulcan reporting marks, are in captive service from the company's quarries elsewhere in Virginia. The railroad uses heavy-duty power, including modern six-axle locomotives, to drag these cars up the steep grade to the industrial park.

There isn't enough trackage in the industrial park to handle all the loads at once, so train crews usually pull the empties first and bring them to a small yard to the east near Van Dorn Street. They then return with the loads. To spot the loads, the engines run around the train using the siding in front of the printing plant. Then they shove the whole cut of cars up the hill into the Vulcan yard.

These trains usually include only Vulcan traffic; cars destined for other businesses are handled in a separate run.

Woodward & Lothrop maintained a large warehouse with a rail-served dock at the southeast corner of the industrial park. Woodward & Lothrop was a department store retailer with a long history in the D.C. area. The warehouse

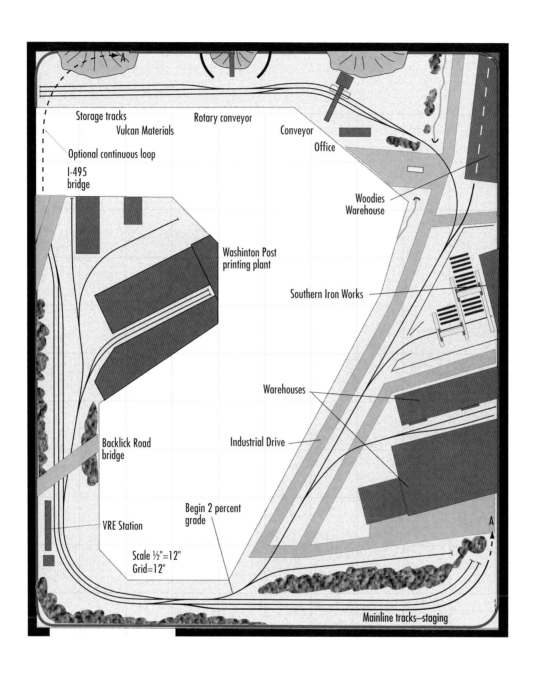

Storage tracks

Vulcan Materials

Rotary conveyor

Conveyor

Office

Optional continuous loop

I-495 bridge

Woodies Warehouse

Washinton Post printing plant

Southern Iron Works

Warehouses

Industrial Drive

Backlick Road bridge

VRE Station

Begin 2 percent grade

Scale ½"=12"
Grid=12"

Mainline tracks—staging

A

LAYOUT AT A GLANCE

Overall size: 10 x 12 feet
Length of line: 28 feet visible
Locale: Springfield, Va.
Period: 1980-1991
Layout style: walkaround, point-to-point
Layout height: 57"
Benchwork: sectional open-grid

Track: code 55
Turnouts: Atlas No. 6, handlaid curved turnouts
Minimum radius: 18"
Maximum grade: 1 percent
Backdrop: .060" styrene sheet
Control system: wireless DCC, hand-thrown turnouts

received large bulk items, such as furniture and appliances, in boxcars. The company went bankrupt in 1995 and closed its doors soon thereafter. The warehouse has since been converted to other non-rail-served businesses, but to maximize the operating potential of the layouts, we can assume that the warehouse is still active and still served by rail.

HO track plans

The two HO plans are very similar in scope. Their main difference is that the first plan is a point-to-point style layout, while the second includes a loop for continuous running. Both concentrate on the southeast portion of the industrial district where Vulcan, Southern Iron Works, and the warehouses are located. The track plan is very faithful to the prototype arrangement. Nearly every siding and turnout is included, although lengths and distances have been compressed. There wasn't enough room to include the *Washington Post* plant in this plan.

There are four stub-end sidings to switch in addition to the Vulcan Materials yard. One or two operators can run the layout, as the central pit is quite roomy. One engineer can run the NS gravel trains and the Vulcan switcher, and the other operator can switch the warehouses and Southern Iron. Usually this train has a set of older, four-axle diesels.

A caboose is frequently used. The layout could also work well with just one operator, with the switching done in a sequential manner.

The second HO plan includes a continuous-run option to provide through trains that use the main line without entering the industrial park, such as Amtrak, VRE, or the coal locals. These trains can run "interference" while the freights try to work the switching district.

The warehouse in the lower right corner is quite large and presents an access problem. A removable access hatch would be useful in this corner. As in the prototype, trees help obscure the main line – in our case, as it ducks under the warehouses. The rest of the layout is readily accessible from the front aisle.

Because there's no runaround track on the layout, except in the Vulcan yard, switching the other businesses will be done with the engine on the right hand side of the train. The staging tracks include crossovers at the far ends to allow engines to run around their trains while in staging. The Vulcan switcher can utilize the runaround and engine pocket on its property to simplify switch moves.

To maximize switching potential and add operational realism, each warehouse should have numbered spots for car delivery. Waybills should specify the exact door where incoming cars must be spotted. In this way a simple stub siding can become two or more switch spots.

Access to the room is best made from the lower left, where a duckunder or removable section can be incorporated into the construction. Layout height is not critical other than ensuring that the Vulcan storage yard clears the staging tracks. Because of the duckunder access, higher is better.

N scale plan

The N scale plan includes the printing plant, which nearly doubles the complexity of operations. It also provides some interesting structures to model, including a section of the Capital Beltway.

The industries on the right side of the layout are the same size as in the HO plan. This results in less selective compression, which will make the buildings look more realistic. It also provides more switching spots for the warehouse sidings.

Two modern NS General Electric Dash-9 diesels shove a long cut of Vulcan cars past the Southern Iron Works.

Several empty boxcars await the next switch job at the printing plant, with the remodeled face of the Robinson terminal behind the cars. Note the curved wall at right.

Warehouses look almost like canyon walls in the industrial park. Foliage, pallets, junk, and structure details all deserve careful modeling.

The Vulcan yard has an interesting variety of conveyors, bins, and heavy equipment for handling aggregates. All would be interesting to model.

The optional continuous run makes more sense here than in the HO plan as the VRE station is modeled, giving commuter trains a place to stop. The double-ended siding in front of the *Washington Post* plant is an important feature. This is where locomotives will run around their trains for the push up the hill to the industrial park.

Access to the inside of the N scale plan is easiest from the upper left where the optional (or removable) continuous run track is located. Another option is to duck under the benchwork from the lower left. Again, layout height isn't critical, but taller layouts require less ducking.

The Robinson Terminal plays a big role in warehousing and storing the newsprint until it is needed, so switching cars from the Robinson Warehouse to the Post plant is an option during operation.

CHAPTER 4

1940 THROUGH 1952

Chesapeake & Ohio's Chicago Division

The Chesapeake & Ohio's Chicago Subdivision crossed the Wabash at CW Tower in Peru, Ind. Here train 93 heads to Chicago behind an A-B-A set of EMD F7s. *Gene Huddleston*

For many modelers, mention of the Chesapeake & Ohio provokes thoughts of long strings of coal hoppers pulled by massive steam engines through the verdant Appalachians. However, the C&O was much more than that. Although much of the railroad was an efficient and profitable double-tracked coal conveyor, the C&O also hauled passengers between New York, Cincinnati, Detroit, and Chicago in famous, named trains and had a multitude of single-track main and branch lines throughout the Midwest.

Rail lines in Indiana, 1947

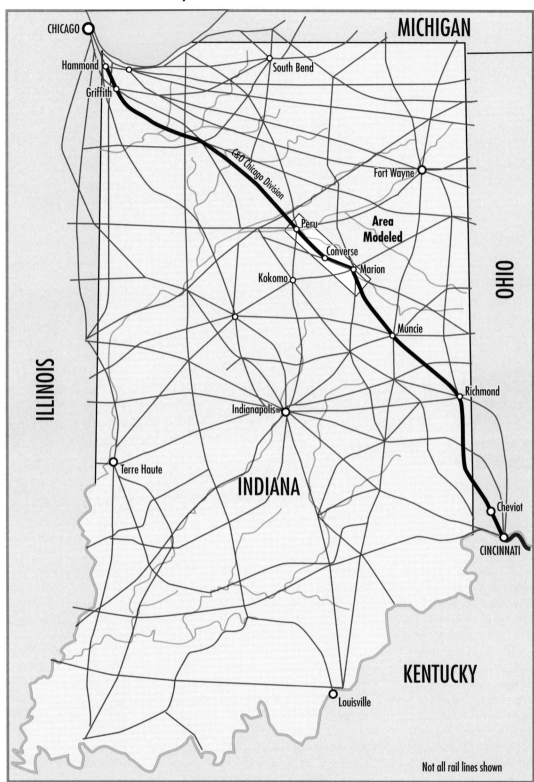

CHICAGO
MICHIGAN
Hammond
Griffith
South Bend
C&O Chicago Division
Fort Wayne
Peru
Area Modeled
Converse
Marion
Kokomo
OHIO
Muncie
Richmond
ILLINOIS
Indianapolis
Cheviot
Terre Haute
CINCINNATI
INDIANA
KENTUCKY
Louisville
Not all rail lines shown

History

The Chicago Division is a little-known and often overlooked part of the C&O system. It has many attributes that make it worth modeling, including hot through freight trains running under timetable-and-train-order authority, interchanges with several railroads, switching opportunities at several manageable-sized industries, some heavy grades and tight curves, numerous interesting bridges, and surprisingly varied scenery. These features can be captured on a layout without being overwhelming for a mid-sized space.

The C&O's Chicago Division traces its roots to the Cincinnati, Richmond & Muncie. By the time that railroad started laying track in 1900, central Indiana was already crisscrossed by an intricate web of railroads (see above). The 285-mile line, finished in 1907 by its successor, the Chi-

Number 1169, a K-2 Mikado, simmers under the coal dock at Peru. The sanding tower can be seen at left behind the locomotive. *Photo courtesy C&O Historical Society*

cago, Cincinnati & Louisville, ran diagonally across Indiana from Hammond, near Chicago, through Peru, Marion, Muncie, and Richmond, then across the Ohio state line to Cincinnati. Because it was built to provide the shortest route between its terminal cities, the railroad advertised itself as the "Short Line."

The CC&L eventually built a line into downtown Cincinnati, but never made it to downtown Chicago on its own rails. Thus the CC&L had to resort to a series of trackage rights and agreements with other railroads to reach the downtown stations and markets of Chicago. In 1906 and '07, the CC&L built a 10-mile line parallel to the Erie Railroad from Griffith, Ind., to Hammond. The two railroads shared their single-track lines, creating in effect a short double-track section.

Also in 1907, the railroad laid track from the state line at Hammond to a

connection with the Chicago Terminal Transfer Railroad and the Illinois Central at Dolton, Ill. The CC&L was then able to use the IC terminal and yard facilities and its Central Station downtown. The first passenger train ran from Chicago's Central Station to Cincinnati's Grand Central Station on April 6, 1907.

In 1910, the consistently profitable Chesapeake & Ohio gained access to Chicago by buying the CC&L. The C&O operated it as a separate entity, the Chesapeake & Ohio of Indiana, under lease until 1934. At that time the C&O fully absorbed the line, making it the Chicago Division.

In the 1940s, the Chicago Division had two subdivisions, divided at Peru. North of Peru was the Wabash Subdivision and south was the Miami Subdivision, both named after rivers. Peru, in the Wabash Subdivision, was the division

headquarters and served as the crew change point. The layout plans in this chapter focus on the line from Peru to Marion.

The C&O expanded and improved the CC&L's engine terminal and repair shop at the west end of Peru. The map on page 37 shows the C&O's tracks in Peru around 1940. There were also minor yards and engine servicing facilities at Cheviot, Ohio, and Hammond, but all repairs to locomotives and rolling stock on the division took place at Peru. In spite of the large engine terminal and shops, the yard at Peru is of manageable size and well suited to a model railroad.

The Wabash was the second major player in Peru. Its main east-west route between Toledo and St. Louis, the route of the *Wabash Cannon Ball*, crossed the Nickel Plate Road at Peru, and their rights-of-way paralleled through town. Both railroads had yards and shops at Peru.

Time freights and interchanges

Chicago Division operations in the 1940s consisted primarily of freight trains, with coal loads making up the majority of westbound traffic. Passenger traffic was limited to a daily two-car local each way between Cincinnati and Hammond.

The chart on page 39 summarizes the number of cars interchanged with other railroads on the Chicago Division. Since Chicago was the end point of the C&O, all cars there had to be interchanged with other railroads or delivered to on-line customers. The chart doesn't include empty cars, though it is reasonable to assume that, on average, an equal number of empties were interchanged.

In Peru, the C&O interchanged with both the Nickel Plate and Wabash. The C&O crossed the NKP at grade just north of Nickel Plate's Wabash River bridge. The Wabash crossing was less than a mile to the west with interchange tracks in the acute angle between the roads. Here the C&O erected one of its standard wooden cabins to guard the crossing (see page 34). On average, the C&O delivered 15 cars a day to the Wabash and 10 cars to the Nickel Plate, with about 40 percent being coal cars. The C&O received seven cars from the Wabash and two from the Nickel Plate on an average day.

The C&O's relationship with the Nickel Plate in Peru went well beyond simple interchange. In 1923 the Van Sweringen brothers gained control of the Nickel Plate, C&O, Pere Marquette, and Erie through a series of holding companies and interconnected directorships. As a cost-saving measure they entered a joint terminal facilities agreement that allowed the two roads to share use of the C&O yard, freight house, and engine facilities at Peru. The C&O maintained the yard and engine facility on the north bank of the river, where Nickel Plate was the tenant. The agreement must have worked well because it lasted until 1951 when the Nickel Plate dieselized its line, well after the Van Sweringen empire collapsed in the Depression.

Just west of Marion, at Phoenix, the C&O had a connection, but did not normally interchange, with the Pittsburgh, Cincinnati, Chicago & St. Louis Railroad, commonly called the Panhandle

C&O, Nickel Plate, and Wabash tracks and facilities in Peru, Ind., circa 1947

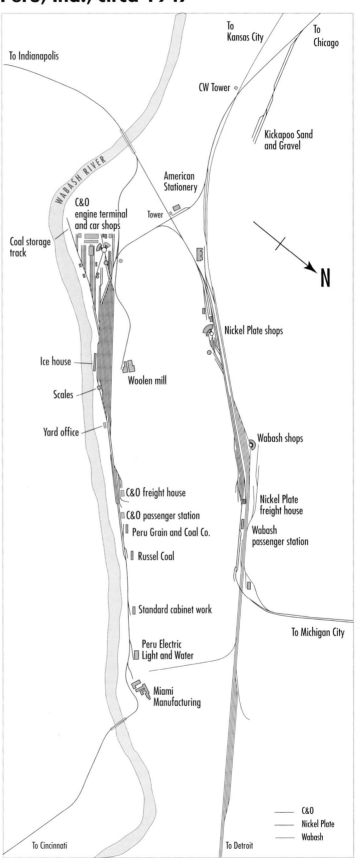

Triple-headed Mikados (2-8-2s) lead a World War II special charity train across the Wabash River. *Photo courtesy C&O Historical Society*

This view of the east yard ladder at Peru shows a diesel switcher and several C&O standard sheds at right. *Photo courtesy C&O Historical Society*

Route. This railroad formed part of the Pennsylvania Railroad system. Its common name came from its main line west from Pittsburgh across the northern panhandle of West Virginia. The line continued west to Bradford, Ohio, where it split, with one line to Chicago and the other to East St. Louis, Ill., via Indianapolis. It paralleled the C&O from Marion to Amboy, and crossed over the C&O on a deck girder bridge near Converse, with the C&O in a cut below.

The C&O interchanged with the Nickel Plate at Phoenix, with the two railroads jointly owning one of the interchange tracks. There the C&O delivered an average of eight cars per day and received five.

Several scheduled time freights carried this traffic each day, including the celebrated *Expediter*. Inaugurated in April 1947, three different sections of this train departed Chicago every weeknight. They originated at Chicago Union Stockyard and Burnham Yard south of Chicago. Cars from the stockyards, Clearing Yard, Calumet Yard, and Red Ball manifests (so called because the composition of these scheduled freights was known or made "manifest" to the receiving yard before they pulled in) from Missouri, Michigan, and Wisconsin destined for the east were assembled into three trains for the C&O. The first section, US-96 on the timetable, hauled only meat. It was scheduled to depart from the stockyards at 5 p.m. every night except Sunday for a 5 a.m. arrival at Cheviot Yard.

The second section of the *Expediter*, CY-90, was scheduled out of the stockyards at 7 p.m. and due into Calumet Yard at 8 p.m. for the Belt Railway of Chicago.

The third and last section of the *Expediter* was Advance 90, leaving Burnham Yard every night at 9:30 p.m. This was an early enough departure to ensure interchange, out of Cheviot, of cars for the Red Ball manifests of the Southern and

Daily average of loaded cars interchanged by the C&O on the Chicago Division

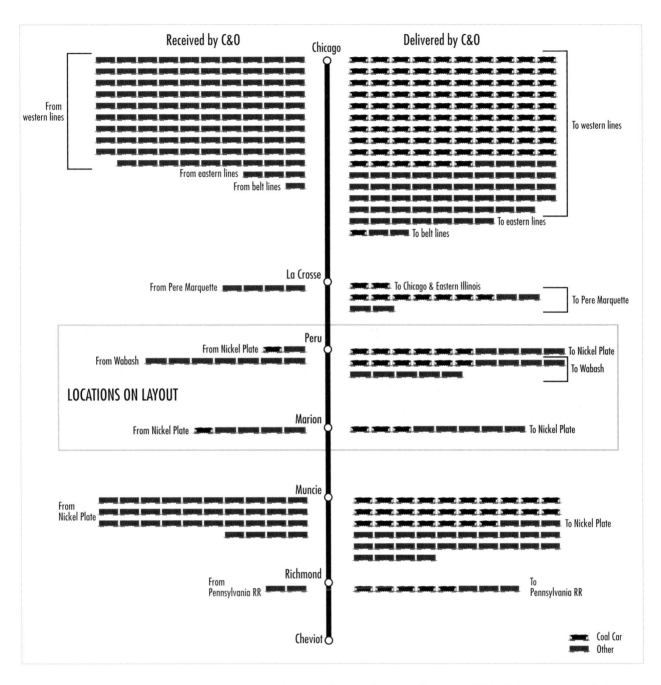

the Louisville and Nashville, which departed from Cincinnati before early afternoon of the next day.

Despite its earlier departure from the Union Stockyards, CY-90 trailed Advance 90 down the Chicago Division due to time for yard work and interchange of cars in Calumet. Thus, while CY-90 halted at Calumet, Advance 90 leap-frogged CY-90 and preceded it to Cheviot.

At Peru the pause to change engines and crews was brief. Even before the train

came to a halt, the flagman had cut off the caboose, and a waiting switchman threw the switch to guide it to an adjacent track. A yard switcher tacked the new caboose to the end of the train, and the new crew grabbed the conductor's paperwork. At the head end, a hostler took the inbound engine to the roundhouse, while the outbound road engine or engines moved into place and coupled to the train. With two blasts of the whistle, the trains started, crossed the bridge over

the Wabash River, and attacked the grade on the other side.

These hot trains did not set out or pick up cars at Peru. That task was left to other freights such as No. 98, a pick-up freight for local trains such as the Peru-Hammond Turn.

Extensive switching of the time freights took place at Stevens Yard, just east of Cincinnati, to block them for future destinations, such as cars that required re-icing at Clifton Forge; cars

C&O Chicago Division, HO scale "mushroom," circa 1940s

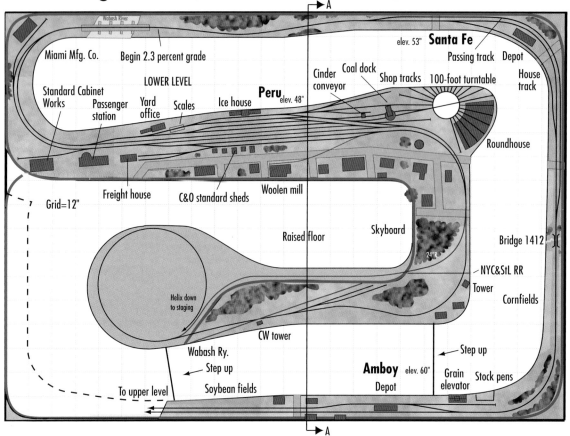

A

Wabash River

Miami Mfg. Co.   Begin 2.3 percent grade

elev. 53"   **Santa Fe**

Passing track   Depot

LOWER LEVEL

Cinder conveyor   Coal dock   Shop tracks   100-foot turntable   House track

Standard Cabinet Works   **Peru** elev. 48"

Passenger station   Yard office   Scales   Ice house

Roundhouse

Freight house   C&O standard sheds   Woolen mill

Grid=12"

Raised floor   Skyboard

Bridge 1412   )(

24"R

NYC&StL RR

Tower

Cornfields

Helix down to staging

CW tower

Wabash Ry.

Step up

Step up

Amboy   elev. 60"   Grain elevator   Stock pens

To upper level   Soybean fields   Depot

A

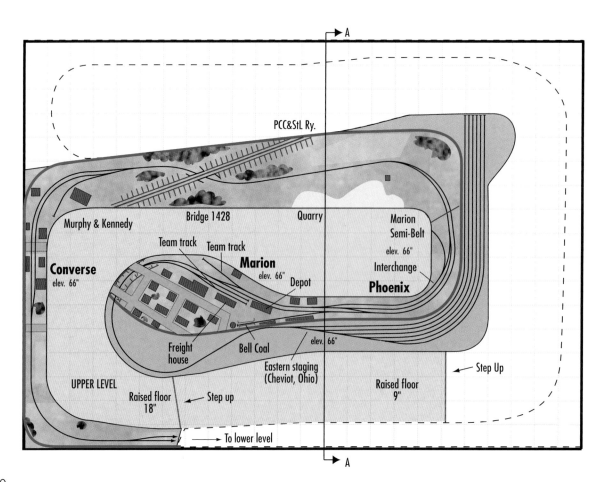

A

PCC&StL Ry.

Bridge 1428   Quarry   Marion Semi-Belt

Murphy & Kennedy   elev. 66"

Team track   Team track   **Marion**   Interchange

Converse   elev. 66"   elev. 66"   Depot   **Phoenix**

Freight house   Bell Coal   elev. 66"

Eastern staging (Cheviot, Ohio)   Step Up

UPPER LEVEL   Raised floor 9"

Raised floor 18"   Step up

To lower level

A

Cross section AA showing "mushroom" portion

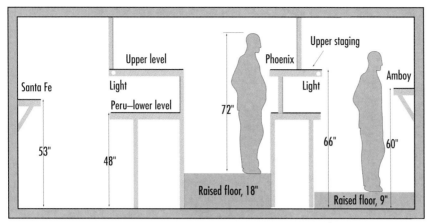

Cross section AA showing mushroom portion of layout

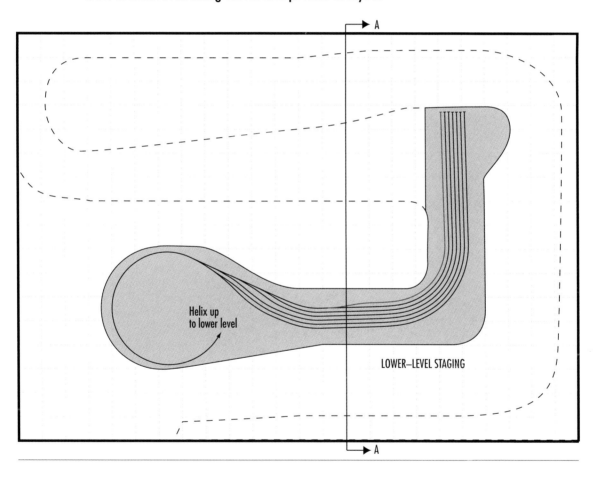

LAYOUT AT A GLANCE

Overall size: 17 x 24 feet
Length of main line: 144 feet
Locale: Miami County, Ind.
Layout style: point-to-point mushroom
Layout height: 48" - 66"
Benchwork: sectional open-grid
Roadbed: cork on ½" plywood, cookie-cutter style

Track: Atlas code 83
Turnouts: Atlas No. 6, handlaid curved turnouts and crossovers
Minimum radius: 30" main line, 24" NKP junction track,
    18" Converse industry track
Maximum grade: 2.3 percent main line, 2.2 percent in helixes
Backdrop: .060" styrene sheet
Control system: wireless DCC, hand-thrown turnouts

A class G-7S 2-8-0 Consolidation leads a freight train near Converse, Ind. The terrain near Converse is flatter than the area near the river. *Photo courtesy C&O Historical Society*

Pick-up freight No. 98 climbs out of the Wabash River valley near Santa Fe, Ind., behind an F7 and a GP7. Note the trees and foliage on either side of the right-of-way. *Eugene Huddleston*

Structures

Since the C&O did not originally build the line, most of the stations were not C&O neo-colonial standards. The Peru, Marion, Richmond, and Muncie depots were brick with large hipped roofs, ornate dormers, and curved extensions. Smaller stations were usually made of wood and built to CC&L and C&O of Indiana standard plans.

Locomotives

Due to the light construction of many bridges and the tight clearances in Chicago, the C&O's famous superpower steam locomotives, such as the class H-8 2-6-6-6 Alleghenies, K-4 2-8-4 Kanawhas, and J-series Greenbrier 4-8-4s did not run on the Chicago Division.

Instead, the division hosted lighter steam engines including a group of 24 booster-equipped K-2 2-8-2s (Mikados), the heaviest locomotives its bridges could handle. One K-4 Kanawha could have handled the load shown at the top of page 38, but instead triple-headed K-2 Mikados had to take this manifest freight across the Wabash River. The G-class 2-8-0s, such as G-7S No. 994 (at upper left), were also common. Class A-16 Atlantics (4-4-2s) and the occasional F-15 Pacific (4-6-2) worked the few passenger and work supply trains.

Due to a World War II motive power shortage, a J-1 Mountain (2-10-4) and two A-16s Atlantics served on the Chicago Division in unusual ways. The J-1 worked the Peru-Hammond turn, releasing a K-2 for much more important work. The two Atlantics were used to double-head 5,000-ton coal trains from Peru to Chicago, with a G-7 helper for part of the way to Fulton.

The Chicago Division was the first part of the old C&O to dieselize. By mid-1952 enough EMD F7s had been received to dieselize everything west of Russell. In the early diesel era, F7 A-B-A sets and GP7s were common.

HO layout designs

The two HO scale layout designs explore the possibilities of modeling the C&O "Straight Line" in multiple decks. Although more complex to design and build, multi-deck layouts allow lengthening the modeled main line. While almost

for interchange for the Southern, Atlantic Coast Line, and Seaboard Air Line at various locations; and cars headed to Newport News and Norfolk. Before that job was completed, a transfer brought the section of CY-90 that had been bypassed on the Chicago Division. This train was pushed over the yard's hump and consolidated with other cars for the race to the "Bay." Thus the rest of the C&O saw just one section of the *Expediter*, with only the Chicago Division seeing three scheduled sections.

Other eastbound freights included No. 92, which departed at 8 a.m., and No. 94,

which left at 9 p.m. On May 16, 1951, the *Expediter* got a westbound counterpart in train 91, the *Speedwest*. Other westbound time freights included Nos. 93 and 95.

The C&O continued to host two symbol freights each way as late as August 1986. By then they were Nos. 300/301 and 508/509.

There was never much passenger traffic on the Chicago Division, and all passenger service was eliminated in the fall of 1949. It's somewhat ironic that Amtrak ran trains on much of the C&O Indiana route from 1974 to 1986.

always a good thing, longer mainline running is especially important to layouts that utilize timetable-and-train-order operations. The longer main line allows more countryside between towns, longer sidings, and thus longer trains. The longer run allows scheduled times between stations to be more realistic, although still compressed, and reduces the chances that a train working one siding will foul adjacent sidings.

In designing a multiple-deck layout, the critical design parameter is the height of each level. In a conventional multiple level plan, it isn't possible to have both decks at an optimum height. The so-called "mushroom" design limits this problem by stacking decks atop each other so that only one deck is visible to the operator at a time. By using a raised floor in the mushroom portion, the levels can be near their optimum heights.

This design also heightens the feeling of isolation between locations and mentally extends the layout, even if the main line is actually shorter than alternate plans. I experienced this effect when operating on the late Jerry Belina's mountainous West Virginia Western, an extremely innovative and effective mushroom layout that was documented in *Model Railroad Planning* in 2005.

The problem with modeling a double-deck railroad in the flatlands of Indiana is disguising the grade between levels. Although most of Indiana is flat and long tangents are common, there are places where small hills and river valleys break up the monotony. The section between Peru and Marion is a case in point. The C&O main line winds and climbs its way out of the Wabash Valley on a 1 percent grade for several miles. Even on the plateau between Peru and Marion, the profile resembles a sawtooth with grades up to 1 percent common.

Let's compare the mushroom and conventional double deck plans. Both fit in a 17 x 24-foot space, and in both cases, the minimum separation between decks is 18". Allowing for 3" of front fascia on the upper deck creates about 15" of usable clearance between decks, which is good on an HO layout.

The aisles are as narrow as possible while maximizing the length of the main line and providing ample people space.

Much of the C&O right-of-way in Marion, Ind., was built on embankments, resulting in many bridges and viaducts. *Photo courtesy C&O Historical Society*

This aerial view of the west end of Peru Yard shows the busy car repair shops. The C&O main to Chicago is at lower left. *Photo courtesy C&O Historical Society*

Thirty-six inches is typical for the mushroom with more room allocated to aisles adjacent to busy areas such as yards. The conventional double-deck layout requires wider aisles, with 48" being typical, because the chances of more than one operator being in the same area are increased. With wider aisles the double-deck plan could not utilize the same "Circle-J" shaped footprint as the mushroom plan with its 36" aisles.

In both plans, the aisles narrow down to less than the typical width for short distances near the helices. This is generally not a problem, as there's room for people to pass on each side of the constriction. The tight sections actually help break the conventional double-deck plan into separate areas, increasing the illusion of distance.

Both layouts cover the same prototype areas, from CW Tower in the west to Marion in the east. Both plans include the same major features (Peru yard as well as several towns, industries, and interchanges). The conventional double-deck plan has 204 feet of main line compared to the 144 feet of the mushroom, not counting the track in the helix between levels. This is its primary advantage.

The hidden track in the conventional double-deck helix is both an advantage and a disadvantage. Four 30"-radius loops at a 2.2 percent grade is 68 feet of track,

Conventional HO scale plan

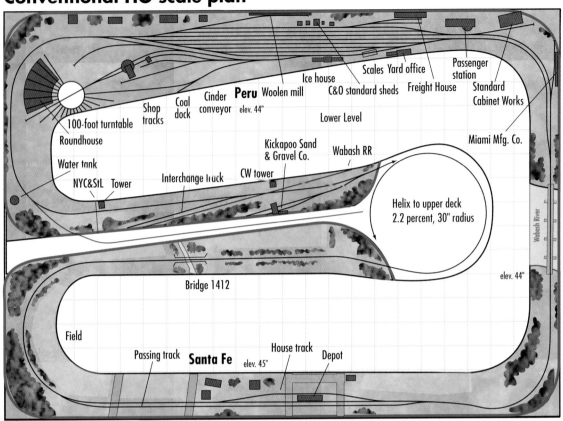

Peru
elev. 44"

100-foot turntable
Roundhouse
Water tank
NYC&StL   Tower

Shop tracks
Coal dock
Cinder conveyor

Woolen mill
Ice house

Scales   Yard office
C&O standard sheds
Freight House

Passenger station
Standard Cabinet Works

Lower Level

Miami Mfg. Co.

Kickapoo Sand & Gravel Co.
Wabash RR
Interchange track
CW tower

Helix to upper deck
2.2 percent, 30" radius

Wabash River

elev. 44"

Bridge 1412

Field

Passing track   Santa Fe
elev. 45"

House track

Depot

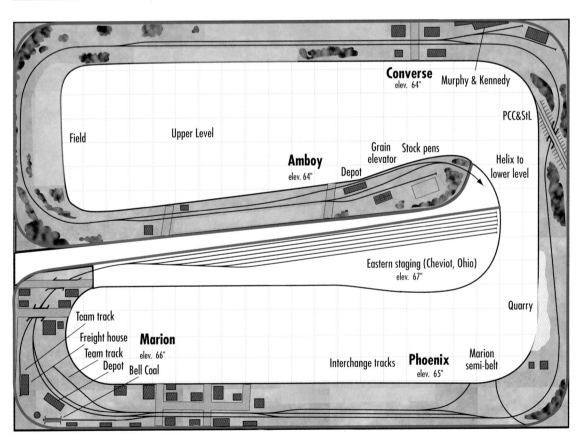

Field

Upper Level

Converse
elev. 64"
Murphy & Kennedy

PCC&StL

Grain elevator   Stock pens

Amboy
elev. 64"
Depot

Helix to lower level

Eastern staging (Cheviot, Ohio)
elev. 67"

Quarry

Team track
Freight house
Team track
Depot   Bell Coal

Marion
elev. 66"

Interchange tracks   Phoenix
elev. 65"

Marion semi-belt

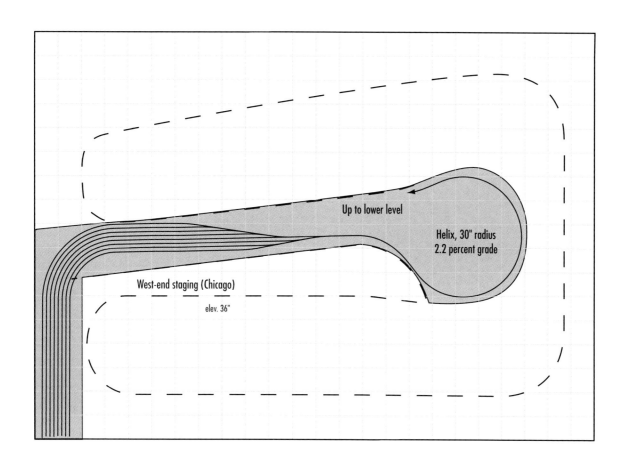

Up to lower level

Helix, 30" radius
2.2 percent grade

West-end staging (Chicago)

elev. 36"

LAYOUT AT A GLANCE

Overall size: 17 x 24 feet
Length of main line: 204 feet (visible)
Locale: Miami County, Ind.
Layout style: point-to-point, double-deck
Layout height: 44" - 66"
Benchwork: sectional open-grid
Roadbed: cork on ½" plywood, cookie-cutter style

Track: Atlas code 83
Turnouts: Atlas No. 6, handlaid curved turnouts
Minimum radius: 30" main line, 24" NKP junction track
Maximum grade: 1 percent main line, 2.2 percent in helixes
Backdrop: .060" styrene sheet
Control system: wireless DCC, hand-thrown turnouts

or about 25 percent of the total 272-foot main line when counting the helix. Although this provides extended running time between towns, operators must wait a considerable time while their trains are out of sight. Since most model railroaders tend to dislike hidden track, this is a disadvantage.

You can alleviate the problem by providing a high-tech closed-circuit television or low-tech peepholes into the helix so that operators can track the progress of trains. Sound-equipped locomotives also help, as they provide clues to the train's

progress. It might be possible to pull one of the loops out of the main helix to make it visible during a portion of the climb, but this will narrow the already tight aisle in that location. If you choose to do this, then the whole central peninsula should be shortened by 6" or so, taking roughly two feet off the overall mainline run.

The mushroom plan doesn't use an enclosed helix to connect the levels. Instead the visible section from the Wabash River to Converse is on a grade. The average grade needed to achieve the 18" deck separation is 2.3 percent. How-

ever, it would be more challenging to operators to vary the grade, with the section from the Wabash River to Santa Fe approaching 3.3 percent and the sections with sidings closer to 1 percent. These grades are steeper than the prototype, but model trains tend to handle grades better that their full-scale counterparts, and the overall effect should be very realistic.

The cross section at AA shows that as the track climbs, so does the raised floor. The drawing shows two distinct steps of nine inches. One is near Amboy and the other near Converse. By using slightly

The C&O crossed the Wabash River east of Peru on a series of deck plate-girder spans. Here FP7 No. 8004 leads the second section of train 98. *Eugene Huddleston*

Alternate design for engine terminal

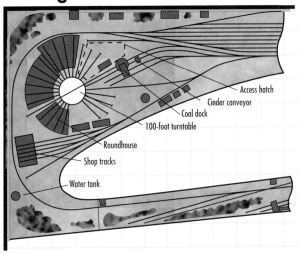

Access hatch
Cinder conveyor
Coal dock
100-foot turntable
Roundhouse
Shop tracks
Water tank

more complex framing, you could build the floor as a ramp with a continuous grade that matches the railroad from the Wabash River to Converse, thus keeping the railroad at the optimum viewing and operating height. This would also lessen the apparent grade to the operator. It also avoids the chance of an operator tripping on a step while concentrating on the moving train.

Both plans use a helix to reach western staging. Modelers generally find using helixes to reach staging not as objectionable as having them on a main line. All staging yards include six tracks with a minimum 12-foot length, the design length for trains on the layout. Passing sidings are at least that length except for

Converse siding on the mushroom plan.

Both layouts are point-to-point, so trains will have to be backed to restage them after operating sessions.

Peru Yard is probably the most important location. Both plans represent slightly simplified versions of the real yard. The double-deck plan devotes the whole length of one wall to this yard, but the curve at the west end is reversed from the prototype. The mushroom plan captures the curve at the west end correctly, but resorts to a 180-degree curve to approach the Wabash River crossing. The conventional double-deck plan does a better job of capturing that curve, although both are sharp compared to the prototype.

The engine terminals in both plans are truncated. They focus on the need to change locomotives and de-emphasize the repair shop functions. The coal dock and roundhouse will be challenging modeling projects. The repair shop was very busy as evidenced by the number of cars visible in the photo on page 43. The alternate engine terminal plan for the conven-

tional double deck expands the terminal to include a better representation of the repair shops, but at the cost of potential access problems to the main line behind it and the west end of Converse siding above. Some type of access hatch will be needed to help with maintenance and construction.

Both plans devote a section of the western staging for Nickel Plate engines that are utilizing the joint terminal agreement and interchange cars. A staging operator/hostler could be assigned the job of bringing trains up from staging to the yard. By adding the hostler tasks, this assignment becomes more interesting than just running trains in and out of staging. A yardmaster can handle the overall yard duties as well as local switching in Peru including the interchanges. The actual breakdown of the jobs can be tweaked once the layout is built and has been operated a few times.

Access to the layout interiors is via a 63" nod-under for the mushroom or a 44" duckunder for the conventional double deck, a decided advantage for the mushroom.

Both plans include tracks to model the interchange between the C&O, Nickel Plate, and Wabash. The mushroom plan places the Phoenix interchanges on a curve, while the conventional double-deck tracks are straight in accordance with the prototype. The interchanges are the primary switching locations, although all towns have sidings and industries that require some switching.

Scenery

Scenic treatment for flatland multi-deck layouts can be tricky. In this case, we benefit by the nature of the terrain in this area of Indiana. Most of the original land had to be cleared of deciduous trees, and many parcels of woodland remain. A line of trees placed along the backdrop will help disguise that joint.

In many cleared areas, farmers planted (or left) trees along the edge of the right-of-way to create wind breaks. This creates the impression that the tracks are cutting through a tunnel in the trees. Including these windbreaks on a layout can break up the prairie look. Several of the prototype photos show how this looked in real life. These types of low, wooded hills are per-

A set of F7s waits in front of the brick station at Peru. The slope in the foreground is the bank of the Wabash River. *Eugene Huddleston*

The Peru yard office was built on pilings over the bank of the Wabash River. Train 92, powered by F units, waits for the yard switcher to finish its duties. *Eugene Huddleston*

There are several bridges, industries, and structures to model. The plan allows plenty of room to model the town of Marion convincingly, particularly in the mushroom plan. Peru will be mainly depicted by painting on the backdrop.

In both layout designs, the lower deck is fully covered by the upper decks. This will require some form of auxiliary layout lighting. Mounting lights along the back edge of the upper-deck front fascia ensures that the cars and structures on the lower deck are not backlit. You could build a valance over the upper deck to provide similar lighting there. With the overall room lights turned off, the lighted decks will draw viewers' attention.

The mushroom layout will require more-complicated benchwork. Cross section AA shows the upper-deck sky board secured to the ceiling. Thick (¼") Masonite hardboard secured to a 1 x 3 wood frame should be sufficient to hold the back of the upper deck secure, since it will also be attached to the sky board of the lower deck. The structural members of the lower deck can be even heavier if desired.

In either variation, the C&O Chicago Division presents an attractive prototype for a mid-sized model layout.

fect for the narrow benchwork that multi-deck layouts entail.

In the flatter sections, such as the stretch of track between Amboy and Marion, we'll need to rely on other techniques to effectively disguise the narrow benchwork. The terrain on the plains is a patchwork pattern of rectangular fields planted with various crops, especially soybeans and corn. Other agriculture includes dairying, egg production, and specialty horticulture. Specialty crops include melons, tomatoes, grapes, and mint. Accurate modeling of these crops and farms will be important.

The general layout of the fields is north to south, while the railroad cuts across at a northwest-to-southeast diagonal. Thus many of the fields will be on a diagonal to the railroad right of way. This can help further disguise the narrowness of the benchwork.

CHAPTER 5

Denver & Rio Grande Western's Craig Branch

Coal is what the Craig Branch is all about. Three Rio Grande EMD Tunnel Motors drag a coal train by Finger Rock, a distinctive scenic element east of Phippsburg. *Paul Dolkos*

Tucked into the northwest corner of Colorado, the Denver & Rio Grande Western's Craig Branch is no sleepy, weed-covered streak of rust through the countryside. In operation since 1914, the branch features a well-engineered right-of-way, heavy rail, Centralized Traffic Control, and 105-car coal trains. It's a branch line with a mainline look and feel that is perfect for a medium-sized pike.

The Denver & Rio Grande Western in Colorado

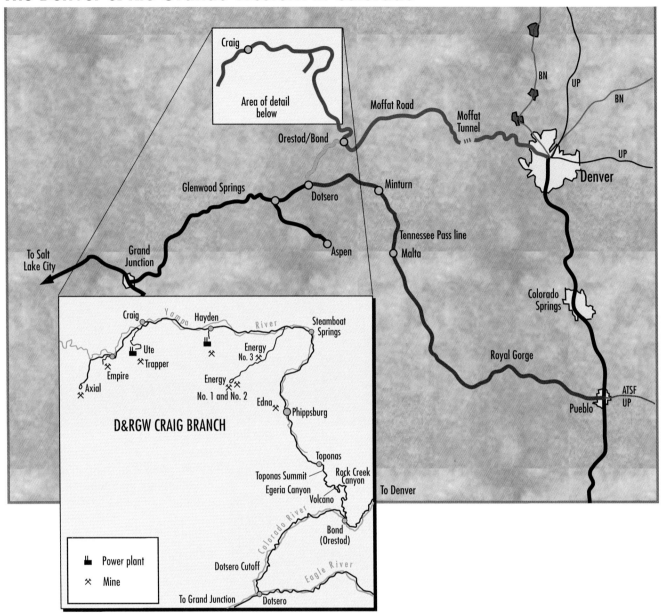

Two summits to cross

This scenic line did not begin its life as a branch. In 1902 the ambitious David H. Moffat, seeking a direct route west from Denver to Salt Lake City, began building a railroad into the Front Range of the Colorado Rockies under the name of the Denver, Northwestern & Pacific Railway.

Backed by Moffat's financing, a 4,000-man crew hacked and blasted a right-of-way with 29 tunnels up South Boulder Canyon. Workers crossed the snow-prone continental divide at Rollins Pass with a series of steep grades, loops, and switchbacks. On the other side of the watershed, the line followed the Fraser River until it reached Yarmony. There construction on

Moffat's road came to a halt as his money ran out.

In 1908, Moffat's friends and local businessmen scraped together $1.5 million to extend the line. But it wasn't easy going. To reach the Yampa River coalfields required an ascent of another mountain, Toponas Summit, with approaches through steep canyons on both slopes. Rock Creek Canyon on the eastern slope was particularly tough. Even the less-confined foothills approaching the canyons required extraordinary construction measures. Engineers had to survey a double horseshoe curve at Crater to gain altitude in order to cross the summit at Toponas. In spite of this, they reached

Steamboat Springs and the Yampa coalfields that same year. With steady coal traffic, the railroad finally started generating some real revenue.

Unfortunately, David Moffat died in 1911 and his railroad went bankrupt soon thereafter. After reorganizing as the Denver & Salt Lake in 1913, the line reached Craig. There construction ceased and the line languished in spite of plentiful coal traffic, partly because the expense of keeping the line open over Rollins Pass in the winter consumed profits. In 1917, the D&SL descended into bankruptcy and remained there until the 1930s.

But the railroad's fortunes would change. In 1927 citizens in Denver and

The loop track for loading coal trains passes under the coal pile at this mine on the Energy branch. *Paul Dolkos*

Four pusher locomotives prepare to shove a loaded coal train out of the yard at Phippsburg. *Paul Dolkos*

Colorado's northwest counties voted $18 million to fund a tunnel under Rollins Pass. The tunnel would also serve as an aqueduct for Denver's water supply while eliminating the D&SL's "stack of steel spaghetti" over Rollins Pass. The railroad portion of the tunnel was completed in 1928 and bears David Moffat's name.

The Rio Grande enviously eyed the tunnel with its associated easier grades and more direct western route from Den-

ver compared to its existing Tennessee Pass route. With some degree of control over the bankrupt D&SL, the D&RGW built a connection from Dotsero on the Tennessee Pass line to Orestod on the Moffat Route in 1934. (Dotsero's name comes from the surveyor's annotation Dot Zero; note that Orestod is Dotsero spelled backward.)

Using the Dotsero cutoff, the Rio Grande began running trains over the

D&SL under trackage rights. The D&RGW formally acquired the D&SL in 1947, giving the railroad a new, shorter western transcontinental route that used the portion of the Moffat Road from Denver to Bond (near Orestod). The remainder of the line from Bond to Craig became a branch. Though Moffat's ambition was finally realized, it did not take the route he had envisioned.

Over the next half century the Rio Grande moved a decent amount of transcontinental traffic, but the bedrock of its revenue was the steady stream of coal that originated on the Craig Branch. Even after two mergers and current Union Pacific control, the Craig Branch still generates three to five coal trains a day.

The Craig Branch hosts a varied sampling of the spectacular scenery for which the D&RGW is famous. Just north of Bond, the line encounters a horseshoe curve at Copper Spur and an overlapping set of horseshoe curves at Crater. Along Rock Creek, the line hugs the steep walls on both sides of the canyon. There are four tunnels in this short section. Trains clinging to both sides of this canyon are an impressive spectacle. The longest sid-

Railroad tracks in Craig, Colo., circa 1999

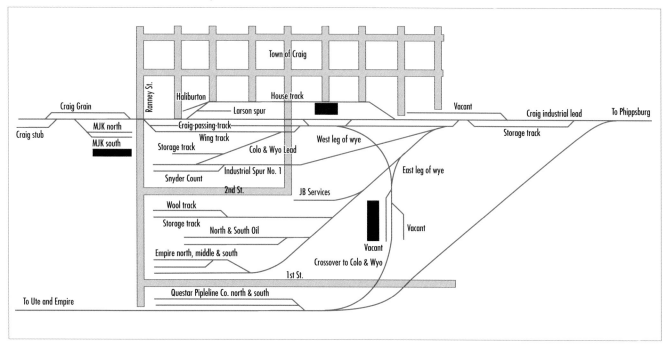

ing on the branch is perched high on the canyon wall at Volcano.

After Tunnel 49, the tracks enter remote Egeria Canyon. Five tunnels later (there are a total of 13 tunnels on the branch), the line reaches the wide-open, grassy expanse of Topanas Summit.

Following Oak Creek, the line enters Phippsburg. Here trains trade Denver-based crews for Phippsburg-based crews, who take trains west to the appropriate mines. Phippsburg hosts a small yard and engine facility, but the town is quite small.

Edna Mine is the reason the line followed Oak Creek instead of the Yampa River. The venerable mine is well known to model railroaders, as it was the inspiration for the Walthers HO and N scale New River Mine kits.

Steamboat Springs was an agricultural center that is now a world-class ski resort. A beer distributor and a lumberyard received rail service in the time period considered in these layouts. The Rio Grande station still stands, though it no longer gets rail service.

The Energy spur reaches an area rich with coal mines. In 1994, the three mines on the spur produced 3 million tons of coal. Over the years several mines opened and closed on this branch.

Farther west, a short spur leads to the power plant at Hayden, although this

Five Rio Grande diesels lead their train over a river near Steamboat Springs. *Paul Dolkos*

plant primarily burns coal trucked in from a nearby mine.

The power plant at Ute consumes more coal than the nearby Trapper Mine can provide, so coal from elsewhere on the Craig Branch is frequently shuttled to this power plant.

The small town of Craig is set amongst the rolling hills and high plains of Colorado's western slope. The railroad

established an intricate set of sidings to serve the ranches and mines in this area. Coal mine services, the gas pipeline industry, Gilsonite mines, grain elevators, and a passenger station provide the basis for an interesting layout design element. The schematic drawing at the top of this page shows the railroad names for the many spurs and sidings in town and some of the industries they served.

Power plant near Craig, Colo., at end of spur

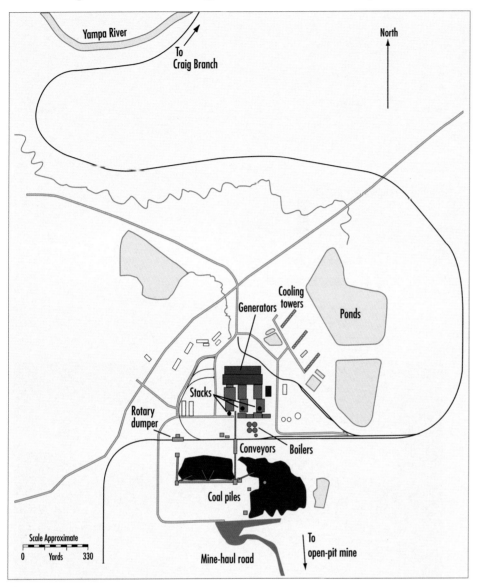

The Axial Spur, the newest part of the line, departs the branch just east of Craig. This spur serves the mine at Empire and the largest mine in the state, Colowyo, at Axial. The power plant also has rail service as shown above.

Operations

Over the years, the traffic on the Craig Branch consisted of moving coal loads east and empties west. Local traffic on the line was always light, ranging from one or two trains a day in the 1950s to a single train per week currently.

Passenger traffic was also light. The *Yampa Valley Mail* ran for many years. In its last years it consisted of one coach and a mail car.

For coal trains, the cycle begins when an empty train arrives at Denver. In 1994, a standard 105-car coal train was assigned six diesels: four on the head end and a pair as a "swing helper" about two thirds back in the train. The train headed to Phippsburg, where the swing helpers were cut out. A fresh crew took the train west for loading and returned to Phippsburg. The swing helper with a second Phippsburg-based crew met the train just west of Phippsburg, where a short grade required the extra help.

At Phippsburg, the Denver-based road crews took over the head end and swing helpers, with a two-unit Phippsburg-based helper tacked on the rear. The train then attacked the 1.8-percent grade to

Topanas Summit. There the Phippsburg helpers dropped off, and the remaining six engines took the train to Denver.

As mentioned above, occasionally coal is shuttled from mines on the branch to the power plant at Ute. The drawing shows the serpentine path the tracks take to access the power plant.

The lighter and shorter Denver to Phippsburg manifest train did not use a swing helper. In 1994, this train ran three times a week. A local Phippsburg crew switched the local industries on the branch with this train.

The layouts

The HO and N scale track plans are each designed for an 11 x 20-foot space. The idea is to use half of a two-car garage. If the layout is located against the outside wall of the garage, an automobile could still be parked in the inner stall. During operation, the car can be moved to provide access to the portions of the layout that face that stall. If built in a spare room or basement, both plans will require an access aisle on one side.

Two N scale plans are shown. The first uses a nod-under with a clearance of about 58 inches from floor to benchwork to allow access to the main portion of the layout. Keeping the nod-under about a foot wide and adding padding under the benchwork will minimize head bashing. This design allows a slightly longer run to Craig and the Axial Branch and keeps the layout 11 feet wide.

The second plan uses an end loop at Craig. This shortens the run by about five feet and increases the overall width of the layout to 13'-4".

The two N scale plans are similar otherwise, and operations on both are the same. Denver staging is under Phippsburg, 48" above the floor. Two operators assigned as Denver head-end and swing-helper crews start their empty train out of staging. While the train is moving west, they walk around the peninsula and meet the train as it emerges from tunnel 44. They follow their train through Rock

D&RGW Craig Branch, HO scale, circa 1980

LAYOUT AT A GLANCE

Overall size: 11 x 20 feet
Length of main line: 44 feet visible, 14 feet
    staging
Locale: Craig Branch, Colo.
Layout style: walkaround oval
Layout height: 54"
Benchwork: sectional open-grid
Roadbed: cork on ½" plywood, cookie-
    cutter style
Track: Atlas code 83
Turnouts: Atlas No. 6, handlaid curved turnouts
Minimum radius: 30"
Maximum grade: none
Backdrop: .060" styrene, mountain horizon
Control system: wireless DCC, hand-thrown
    turnouts, CTC optional

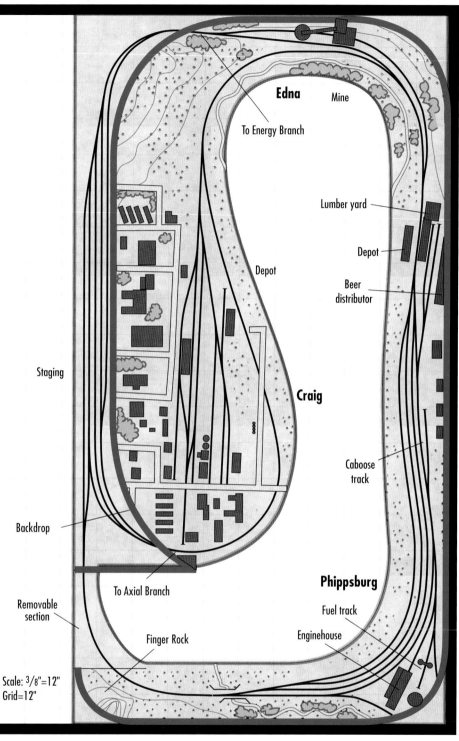

Creek Canyon to Phippsburg, where they would terminate their run as described earlier. Another crew then takes the train to the mine.

Crews heading to Energy diverge onto the branch and into a helix to bring the empty train back to staging. They cut off from their empty train in staging, run around a parked, loaded train, and bring it to Phippsburg. This places an empty train in staging, ready for its next run. The loaded train arrives in Phippsburg where the Denver-based crews take over. Helpers are cut in, and the loads head east to Denver. The grade in the helix is less than the run in Rock Creek Canyon, so helpers shouldn't be necessary to bring the loaded train up to Phippsburg.

The grade in Rock Creek Canyon is reversed from the prototype. This is necessary to allow for hidden staging under the main layout. The helpers heading to Denver are there for braking only.

Coal trains to Axial and Edna will follow a different pattern, basically a loads-in/empties-out shuttle run. At the start of a session, a loaded coal train is waiting at

Craig Branch, N scale, with duckunder

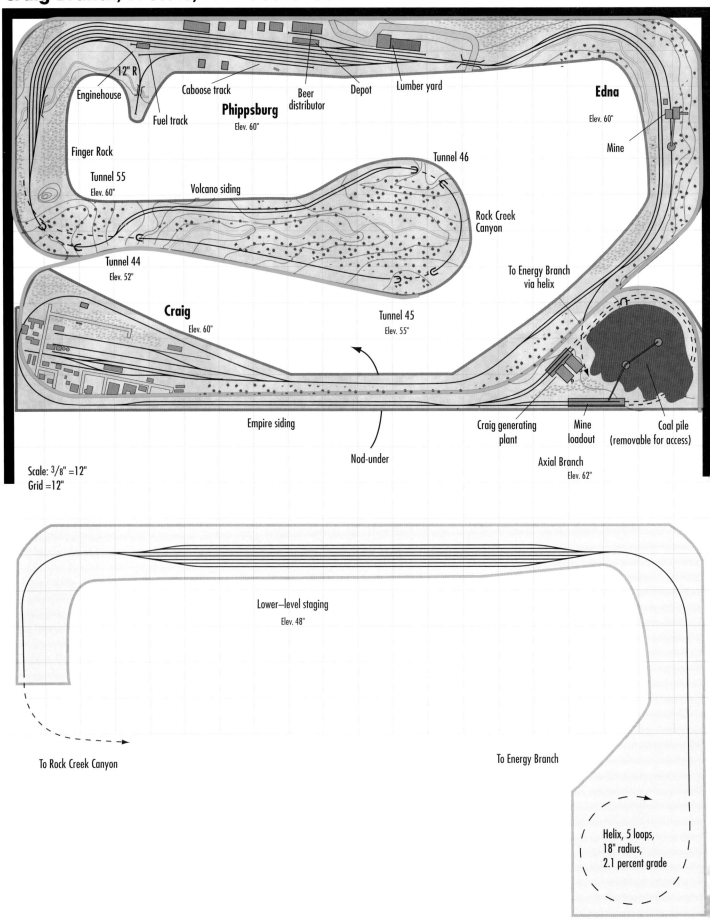

12" R

Enginehouse

Caboose track

Beer
distributor

Depot

Lumber yard

Edna
Elev. 60"

Fuel track

Phippsburg
Elev. 60"

Mine

Finger Rock

Tunnel 46

Tunnel 55
Elev. 60"

Volcano siding

Rock Creek
Canyon

Tunnel 44
Elev. 52"

To Energy Branch
via helix

Craig
Elev. 60"

Tunnel 45
Elev. 55"

Empire siding

Craig generating
plant

Mine
loadout

Coal pile
(removable for access)

Nod-under

Axial Branch
Elev. 62"

Scale: 3/8" =12"
Grid =12"

Lower–level staging
Elev. 48"

To Rock Creek Canyon

To Energy Branch

Helix, 5 loops,
18" radius,
2.1 percent grade

Alternate N scale plan, no duckunder

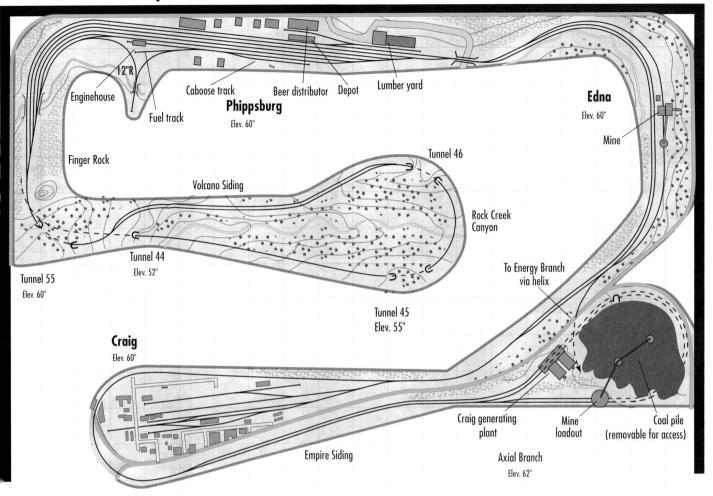

Phippsburg Elev. 60"

Enginehouse
12"R
Fuel track
Finger Rock
Caboose track
Beer distributor
Depot
Lumber yard
Edna Elev. 60"
Mine
Tunnel 46
Volcano Siding
Rock Creek Canyon
Tunnel 44 Elev. 52"
Tunnel 55 Elev. 60"
To Energy Branch via helix
Tunnel 45 Elev. 55"
Craig Elev. 60"
Craig generating plant
Mine loadout
Coal pile (removable for access)
Empire Siding
Axial Branch Elev. 62"

Scale ³/8"=12"
Grid=12"

Alternate plan showing walk around versus nod-under. Staging level is the same.

LAYOUT AT A GLANCE

Overall Size: 11 x 20 feet
Length of main line: 93 feet
Locale: Craig Branch, Colo.
Layout style: walkaround oval
Layout height: 48" - 60"
Benchwork: sectional open-grid
Roadbed: cork on ½" plywood, cookie-cutter style

Track: Atlas code 55
Turnouts: Atlas No. 6, handlaid curved turnouts
Minimum radius: 18", except where designated
Maximum grade: 3.5 percent
Backdrop: .060" styrene sheet, mountain horizon
Control system: wireless DCC, hand-thrown turnouts, CTC optional

The mine tipple at Edna is the prototype for the Walthers New River Mine kit. Here several diesels slowly pull their train through the loading area. *Paul Dolkos*

Edna. The Phippsburg-based shuttle crew will take this train to Axial and park in one of the sidings. A second Phippsburg crew will bring an empty train from Phippsburg to Axial and park in the other siding. Both crews will cut off their power and swap trains. The empty train will head back to Edna, while the loaded train returns to Phippsburg and then Denver using the same pattern of crew swaps as described earlier. At the end of the session, this train is restaged at Edna.

To add variety, each session could also see a mixed freight. A Denver-based crew brings it out of staging, while a local Phippsburg crew does the appropriate switching at Phippsburg and Craig. The industries at the real Steamboat Springs are modeled at Phippsburg to save space. The town of Craig has a more accurate depiction of the tracks and industries requiring service. One could add a track or two at the Craig power plant to provide an additional spot for switching.

Due to the increased space requirements, the HO plan ignores the moun-

A local freight with several flatcar loads of pipe works at the American Gilsonite Co. in Craig. *Mike Danneman*

tain climb and focuses on the operation from Phippsburg to Craig. There's no room for a central peninsula. Staging tracks behind Craig, but accessible from the rest of the garage or aisle when not in a garage, represent both Denver and the

coal mines on the branch. A removable layout section provides a connection for empty trains coming from Denver to Phippsburg. This section can be taken out during construction and maintenance, but must be installed during operation. This will require operators to nod under the section while running trains. The drawing suggests a 54" height for this, but you can adjust the height if desired.

Empty coal trains leave Phippsburg and return to staging via the Energy or Axial branch. Loaded coal trains in staging take the opposite route. A loaded coal train staged in Edna at the start of the session can return to Denver via Phippsburg or shuttle to the Craig power plant via the Axial branch. The Edna mine shares its lead with the Energy branch. Basically, the staging acts as a glorified loads-in/empties-out operation.

The prototype Craig Branch is fully signaled with Centralized Traffic Control. Adding a signal system to any of these layouts will increase operating enjoyment. If no signal system is used, a dispatcher can still be an interesting operating position. Dispatching duties could be combined with a staging operator to orchestrate the swapping of loads and empties.

In the HO plan, Phippsburg is the only place to set up meets between trains,

The spectacular Rock Creek Canyon highlights the scenery found along the Craig Branch, and the area is designed to be a scenic highlight of the N scale track plans. *Mike Danneman*

other than in staging. The N scale plan features four possible locations for meets, giving the dispatcher more options during operations.

A wide variety of motive power is available in both scales to model this branch. One could even backdate the layout to the steam era in HO. There are not many suitable N scale models for modeling the steam era. Backdating would increase the local switching options, especially in Craig.

Two GP30s and a snowplow pause near the sanding towers on a winter day at the small engine-servicing facility at Phippsburg. *Paul Dolkos*

CHAPTER 6

Erie/LTV
Mining Railroad

Steam rises as iron ore pellets drop from hopper cars into the loading bins at LTV's Taconite Harbor, Minn., ore dock. The dock on Lake Superior can handle thousand-foot Great Lakes ore carriers.
John Leopard

The Mesabi Iron Range in northern Minnesota contained the richest deposit of iron ore in the United States, but the nonmagnetic and powdery quality of the soft hematite ore delayed its discovery and exploitation until the 1890s. When its great value became known, there was an unparalleled scramble to exploit it. Nearly all of this high-grade natural iron ore in Minnesota had been mined out by the 1950s. Fortunately, advances in technology found a use for taconite, a lower grade iron ore also plentiful in the region.

In the 1950s, the Erie Mining Company began to mine and transport this processed taconite to Midwestern steel mills. The Erie Mining Railroad was the company's self-contained, efficient operation, which featured a varied locomotive roster and a unique waterfront ore dock. The railroad, mines, and processing plant are excellent candidates for a mid-size layout that will provide many modeling and operational challenges.

From waste rock to industry staple

When high-grade natural iron ore was plentiful, taconite was considered a waste rock. But as the supply of high-grade natural ore decreased, the mining industry began to view taconite as a resource. In the 1930s and '40s, scientists and engineers developed a process to remove the iron ore from the taconite rock and process the material into pellets.

The taconite manufacturing process begins by blasting the very hard taconite rock into small pieces. Giant dump trucks haul the rock to the processing plant.

LTV's northern Minnesota line

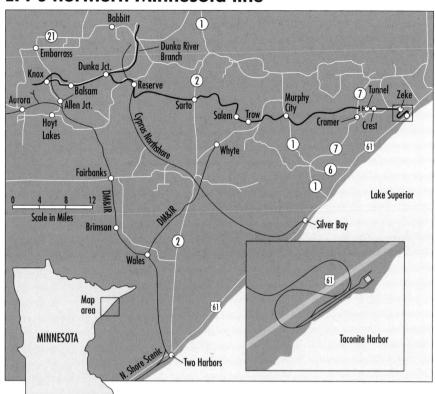

Alco RS-11s usually handled the mine and transfer runs. This large dump truck is one of many hauling raw ore from the Donora pit to the train loading point. *John Leopard*

Overhead washers spray water on the taconite pellets prior to their trip to Taconite Harbor. *John Leopard*

A loaded ore train blasts out of the east portal of Cramer Tunnel. Fall color has started to show, meaning it must be August in northern Minnesota! *John Leopard*

At the plant, the taconite is crushed, mixed with water, and the iron ore in the powder is magnetically separated. The taconite powder with the iron in it is called concentrate. The remaining rock (called tailings) is waste material and is dumped into basins or piles.

The concentrate is mixed and rolled with clay inside large rotating cylinders to form marble-sized balls. These balls are then dried and heated until they are white hot. The balls become hard as they cool. The finished taconite pellets are shipped to steel mills, where blast furnaces begin their conversion into steel.

In 1948 the Erie Mining Company opened a taconite processing plant near Hoyt Lakes. In 1957 the company expanded the plant and built a 74-mile railroad to link Knox Yard near the pellet plant to Taconite Harbor on Lake Superior. LTV Steel Mining absorbed the bankrupt Erie Mining in 1986, but operations remained essentially unchanged until LTV ceased operation in 2001.

During operation, LTV mined about 25 million tons of crude ore each year.

Taconite Harbor

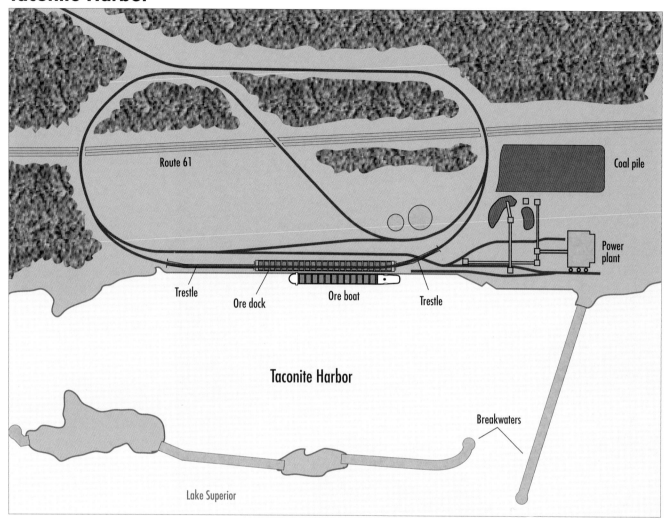

Route 61

Coal pile

Power plant

Trestle

Ore dock

Ore boat

Trestle

Taconite Harbor

Breakwaters

Lake Superior

There were two sets of mines: The primary mines, comprising nine different pits, were near the pellet plant. Some of these were rail served and others used dump trucks. The other site was the Dunka Mine, located about 10 miles down the main line from Knox Yard. Mine-run trains consisting of 18 to 36 Difco or Magor side-dump cars worked these two sites. The trains were equipped with a control cab on the rear end so that a single crewman could operate the train in push-pull manner as it shuttled crude ore to the pellet plant.

The concentrator converted this crude ore into taconite pellets that were about 67 percent iron. Yard switchers moved cuts of ore cars to a tipple to be loaded with pellets. At Knox Yard, the loaded cars were then assembled into trains of 96 to 120 cars. Once the road crew had been called and air brakes tested, the train traveled the 74 miles to Taconite Harbor in

The taconite plant is visible in the distance behind this A-B-B-A set of F9s hauling loads bound for Taconite Harbor. *John Leopard*

LTV/Erie Mining Co., HO scale circa 1980-91

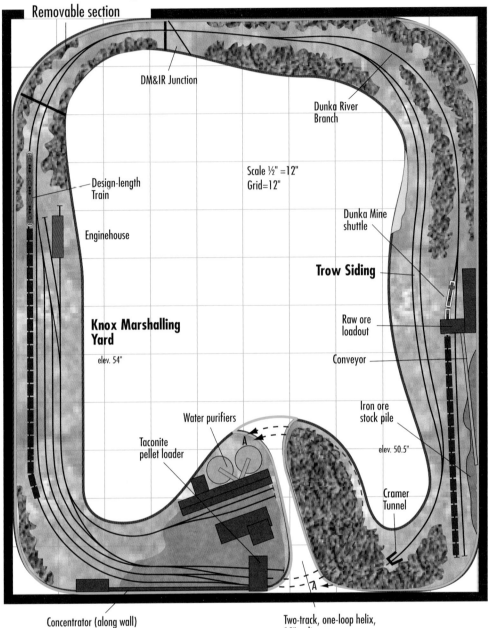

Removable section

DM&IR Junction

Dunka River Branch

Design-length Train

Scale ½" =12'
Grid=12"

Enginehouse

Dunka Mine shuttle

Trow Siding

Knox Marshalling Yard

elev. 54"

Raw ore loadout

Conveyor

Water purifiers

Iron ore stock pile

Taconite pellet loader

elev. 50.5"

Cramer Tunnel

Concentrator (along wall)

Two-track, one-loop helix,
18" radius,
3.2 percent grade

LAYOUT AT A GLANCE

Overall size: 10 x 12 feet
Length of main line: 29 feet (14 feet in helix staging)
Locale: northern Minnesota
Layout style: walkaround oval
Layout height: 54"
Benchwork: sectional open-grid
Track: code 83
Turnouts: Atlas No. 6, handlaid curved turnouts
Minimum radius. 30" main line, helix 18"
Maximum grade: 3.2 percent (in helix)
Backdrop: .060" styrene sheet
Control system: wireless DCC, hand-thrown turnouts

---

about three hours. Several passing sidings along the route allowed multiple trains to operate on the single-track line.

The LTV crossed the Duluth, Missabe & Iron Range and Cyprus Northshore lines, then snaked through rolling hills, wild swamps, and forests as it left the iron range. Train crews frequently spotted moose and other wildlife along the way. The line entered Cramer Tunnel, crossed a high trestle, and passed through a deep rock cut. Three horseshoe curves near Taconite Harbor eased the railroad down a 2 percent grade to the lake.

Taconite Harbor is located in a remote, undeveloped location on Lake Superior 70 miles northeast of Duluth. Here two small islands created a natural breakwater, and the harbor was built in the sheltered water. The Erie/LTV ore dock is unique on the Great Lakes in that it does not jut into the lake as do others, but instead runs parallel to the shore. A steel trestle carries the tracks over the ore storage bins. The tracks loop around and over the bins, allowing trains to unload and begin their return trip without stopping or switching cars.

The ore dock has 25 storage bins on 48-foot centers. The bins have telescoping arms that can extend 91 feet over a ship. These arms can travel back and forth, allowing the ship's holds to be evenly loaded. Multiple arms can transfer ore at once, allowing 58,000 tons of pellets to be loaded in four or five hours. The ore dock and harbor can handle the largest lake carriers, even the newest 1,000-foot-plus ships.

The ore cars had hopper doors attached to a lever mechanism connected to rubber wheels/tires mounted on the car

A front-end loader and crane with clamshell bucket load pellets into waiting ore cars. The rubber tires mounted low on the sides of the cars trigger the hopper door mechanism at the loading dock. Note the red ore dust that weathers most equipment. *Dan Mackey*

sides. As the cars passed slowly over the ore bins, an actuating arm rose and pushed against one of the wheels, opening the hopper doors and releasing the pellets. Another actuator pushed on the opposite wheel to close the doors. This system unloaded a train in about 15 minutes.

At the east end of the ore dock is a 510-foot-long coal dock for a power plant. Erie Mining originally built the plant to supply electricity to the mining operation and community, and LTV sold the plant to Minnesota Power in 2001. This plant receives low-sulfur subbituminous coal by boat.

Motive power

The railroad's varied locomotive roster was a railfan's delight, including EMD F9, GP38, and rebuilt GP20M and GP38-3 diesels; RS-11, C-420, and C-424 engines from Alco; and S-12 switchers from Baldwin. The RS-11s stayed mostly on the mine runs and the S-12s worked Knox Yard. The F9s gained fame

An Alco RS-11 spots loads of raw ore at the processing plant's unloading area. *John Leopard*

as the country's last remaining cab units in regular freight service. It was possible to see matched F9 A-B-B-B-B-A sets on ore trains as late as 2001. Although owned by LTV after 1986, many of the locomo-

tives maintained their Erie Mining Company markings until the end.

In 2001, LTV shut down the aging Hoyt Lakes plant. As a final farewell, in 2004, the new owner, Cliffs-Erie, briefly

LTV/Erie Mining Co., N scale

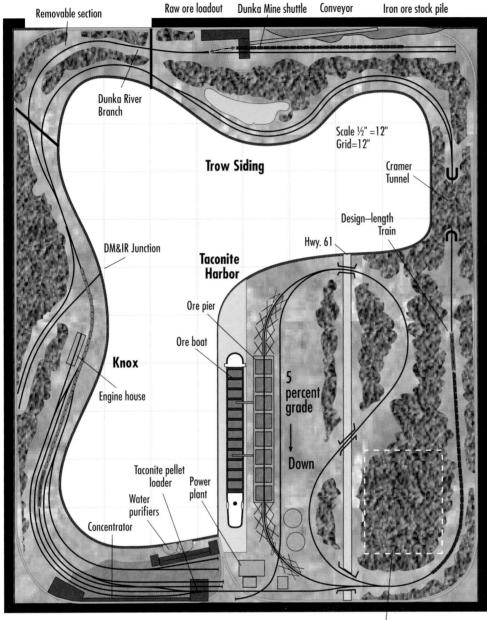

Removable section   Raw ore loadout   Dunka Mine shuttle   Conveyor   Iron ore stock pile

Dunka River Branch

Trow Siding

Scale ½"=12"
Grid=12"

Cramer Tunnel

DM&IR Junction

Taconite Harbor

Design—length Train

Hwy. 61

Ore pier

Ore boat

Knox

Engine house

5 percent grade

Down

Taconite pellet loader

Power plant

Water purifiers

Concentrator

Access hatch

LAYOUT AT A GLANCE

Overall size: 10 x 12 feet
Length of main line: 44 feet
Prototype: LTV/Erie Mining Company
Locale: northern Minnesota
Layout style: walkaround oval
Layout height: 54"
Benchwork: sectional open-grid
Track: code 55
Turnouts: No. 6, handlaid curved turnouts
Minimum radius: 18" main line
Maximum grade: none
Backdrop: .060" styrene sheet
Control system: wireless DCC, hand-thrown turnouts

---

restarted the railroad using a set of surviving F9 A-B-B-A engines in a series of special runs to reclaim pellets and fines that had been crushed into the base earth under the massive storage piles.

The operation's closing left more than 1,400 workers without jobs, a devastating blow to the area economy. The railroad remains intact, and it remains to be seen if a new operator will reopen this fascinating operation.

Track plans

These room-sized track plans depict the mainline haul from Knox Yard to Cramer Tunnel, and the Dunka Branch in N and HO scale. Both plans have one passing siding for meets, although the Dunka Branch can also be used for meets if necessary, a tactic that the prototype railroad used occasionally.

The N scale plan includes a compressed version of the Taconite Harbor loading dock. There wasn't enough room to include this impressive structure in the HO plan. Instead, the HO plan uses a two-track, single-loop helix to represent the ore dock. A loaded train that enters Cramer Tunnel can be stored in the helix. When needed, it can enter Knox Yard

from behind the pellet loader, now representing a loaded train arriving from the local pits. An empty train waiting in the helix can emerge from Cramer Tunnel for the return trip to Knox Yard.

The N scale plan presents a different challenge: how to empty the cars on the ore dock. One solution is to use removable loads that can be taken out while the train is on the trestle. Lightweight cast loads with embedded steel strips or small ball bearings could be removed with the aid of a magnet.

The plan includes an alternative using the connection track from the ore loop through the backdrop to Knox Yard. Once the loaded train traverses the elevated trestle, it would pull into the position shown by the design-length train on the plan. It would then move backward using the connection track to Knox. Once clear, an empty train would use the connection track to emerge by the power plant for the return trip to the yard. This operating scheme isn't ideal, as it involves several non-prototypical moves and a connection track that leads mysteriously into Lake Superior. If neither of these schemes is to your liking, you could use cars with loose loads and operating doors, but as they say in engineering school, "this is an exercise left to the reader."

Don't forget to include the automotive tires mounted on the sides of the ore cars. This is a unique detail that sets the Erie Mining/LTV ore cars apart from others.

Both track plans include highly compressed models of the concentrator and pellet plant. The actual structures are gigantic, nearly two-thirds of a mile long. The plan attempts to simulate the bulk of the structures with flats along the back wall, as well as a few three-dimensional structures arranged in such a way to hide the entrance to the staging loop or connection track. The water treatment settling ponds are prominent aspects of the pellet plant and some representation of them should be included in the model.

One of the best aspects of modeling this railroad is the varied roster of engines. Models of most locomotives are available in both scales, though custom painting will probably be required. The mine locomotives tended to be heavily weathered, while the road engines stayed cleaner.

Both plans include the crossing with the Duluth, Missabe & Iron Range to match the prototype Erie/LTV Mining line's DM&IR interchange near the pellet plant. The N scale plan also includes an interchange track to spice up the operations. One possible use of the interchange is bringing in coal loads for the power plant at Taconite Harbor. Although the power plant received most of its coal by lake boat, it does have rail access. Covered hoppers loaded with clay for the pellet plant and boxcars of explosives are two other possible interchange sources. The DM&IR track comes in from the left

Three F9s spliced by a GP38 lead a string of empty ore cars out the west portal of Cramer Tunnel and back to the processing plant. Area scenery features rolling hills and lots of trees. *John Leopard*

Tracks crisscross at the taconite plant. The DM&IR track enters from left and crosses under the Erie/LTV bridge. The concentrator is at upper left, the pellet plant at center, and the pellet piles at lower right. *Dan Mackey collection*

under the Erie/LTV mine run track. There is a connection track to the left of the pellet plant.

The scenery in this part of Minnesota is heavily forested with both deciduous and coniferous trees. The terrain is gently rolling, with low ridges, one of which is pierced by Cramer Tunnel. This is an unpopulated area and there are no towns along the right-of-way. The line passed several lakes, and one is included in each plan. Wildlife is plentiful; moose are often seen, and a few should be included as part of the layout detailing.

CHAPTER 7

[**1980s**]

Utah Railway

The S curve at Kyune, Utah, is a favorite spot for railfans and photographers along the Utah Railway. Here, four EMD SDs in the railroad's modern gray and red scheme lead a loaded coal train.
Ron Burkhardt

Early travelers struggling westward across the alien landscape of eastern Utah were rewarded with an even more difficult challenge. Rising from the parched desert basin, the sheer face of the Wasatch Mountain Range presented a major obstacle on the way to Salt Lake City and its fertile valleys. The few available passes wind through sheer canyons with loose rock and steep grades. Heavy snow and avalanches often made them impassable. Although a challenge to early settlers and railroad builders, this area presents a wonderful subject for a model railroad.

Utah Railway, 1990

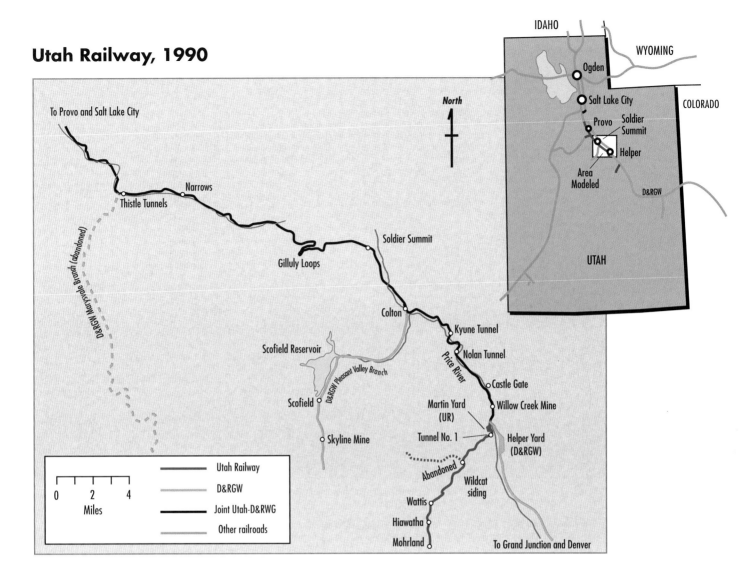

A reluctant partnership

In 1878, the Utah & Pleasant Valley Railroad opened a narrow gauge route westward through the daunting Wasatch Mountains. The line followed the fast-moving Price River almost to its summit. Despite two tunnels and a water-level route, the best the engineers could manage was a steady 2.4 percent westbound grade. Departing the river's shoulder, the route crested the massif at the wind-blown plateau of Soldier Summit, named in honor of a party of homeward-bound Civil War soldiers that perished there one winter. This is Utah's highest (7,440-foot elevation) and steepest main line.

Descending to the west, the grade tilted to more than 4 percent before again following a river, this time the Spanish Fork, to Provo and then north to Salt Lake City. The Denver & Rio Grande Western bought the UP&V in 1882 in its quest for a western outlet and converted the railroad to standard gauge in 1890.

In 1911, A.B. Apperson, general superintendent of the D&RGW's Utah lines, had a dispute with H.U. Mudge, the president of the Rio Grande. As a result of the conflict, Apperson, with backing from New York financiers and Utah mining concerns, proposed building a railroad to bring coal from the rich mines of Carbon County to Provo and a smelter at Midvale, Utah. He incorporated the Utah Coal Railway Co. in January 1912, with the name later amended to the Utah Railway Co.

The company's goal was to build a railroad with an easier route than the torturous D&RGW line. The first proposed alignment, from Spanish Fork through Huntington Canyon to mines at Mohrland and Hiawatha, was impractical, so the railroad instead opted for a line that in many places paralleled the existing D&RGW line, though at a lesser grade.

After seeing the Utah Railway complete the first ten miles of grading of this

line, and mindful of a potential competitive threat from the Union Pacific, the Rio Grande proposed a joint project to the UR. The D&RGW would improve its route with a double-track main line and reduced grades over the summit to Thistle and grant the Utah Railway trackage rights if the Utah Railway would give the D&RGW trackage rights on its new line from Thistle to Provo. In effect this would create a first-class double-track railroad across the vicious Wasatch Mountain range.

By November 1913, after long negotiations, both parties formally signed a joint-use agreement. As part of the bargain the D&RGW created the famous Gilluly Loops. Through this series of adjacent horseshoe curves, engineers were able to reduce the maximum western slope from 4 to 2 percent.

Meanwhile, the Utah Railway completed the construction of the rest of its main line from Panther Junction to

One of Morrison-Knudsen's distinctive MK5000C diesels idles with its nose door open near the Utah Railway enginehouse at Martin, Utah. *Bernard Kempinski*

Hiawatha and Mohrland on October 31, 1914. Several branches radiated out to the numerous coal mines in the area. A small yard at Hiawatha acted as a marshalling yard for loads.

This line includes the impressive Gordon Creek Bridge, one of the longest steel girder bridges in Utah. Still in use, it stands 135 feet tall, 634 feet long, and is built on a curve. The railroad also built a yard at Martin near Helper with an engine terminal, shops, and company headquarters. The D&RGW already had a sizable presence at Helper, which was named after the helper locomotives that worked there.

With the main line and branches complete, the Utah Railway took over the operation of its own line in 1917 after an experiment with D&RGW running mixed freights proved unacceptable. The Rio Grande dispatchers controlled the line. At this point, the UR joined with the San Pedro, Los Angeles & Salt Lake Railroad, now the Union Pacific, to provide a western outlet for traffic. This alliance was allowed in the trackage agreement with the D&RGW. The UP and the UR maintained a close relationship

through much of its subsequent history. Many UR steam locomotives and cabooses were based on UP designs.

With an initial set of four mine runs, the UR began independent operations. The railroad even ran limited passenger service in its early years.

The 1930s brought the Great Depression. Some of the coal mines played out, and towns were abandoned. Several branch lines were abandoned or left idle during this time, but the core business of hauling coal remained. Coal traffic picked up with the onset of World War II.

In 1952, the UR began to dieselize, purchasing six six-axle RSD-4 locomotives from Alco. These white and red striped engines quickly weathered to a light gray, and were magnets for railfans and photographers for nearly 30 years.

Modern operations

The Thistle Tunnels are perhaps the newest railway tunnels in the United States. They were built in response to a massive landslide in 1983 that dammed the river, flooding the town of Thistle and the D&RGW's small yard. In addition, the

slide forever cut off rail access to the Marysvale branch. To avoid the slide site, the railroad and highways were relocated higher up the north slope. To maintain the grade the railroad had to dig two 3,000-foot tunnels.

The year 1985 was a watershed for the UR. That's when cabooses were replaced by flashing rear-end devices, new automated loading facilities at Wattis and Wildcat opened, and the railroad started leasing power and maintenance, first from Helm Leasing, then Morrison-Knudsen, and most recently from Aomni-tracx.

While the UR was steadily hauling coal, its neighbor and partner D&RGW was also busy. For many years, the Soldier Summit route was part of the Rio Grande's main line, providing part of the transcontinental bridge route across Colorado and Utah. In 1981, the D&RGW acquired the Southern Pacific and took on its name. Then in 1996, the Union Pacific and Southern Pacific merged. At that time the Interstate Commerce Commission granted the Burlington Northern Santa Fe trackage rights over the former D&RGW main line, now UP's new cen-

Utah Railway, HO track plan, 1980-93

Power plant

Panther Jct.
elev. 48"

D&RGW
staging tracks

Castle
Gate Coal

Gordon Creek Trestle

Hiawatha
elev. 49"

Load out

Enginehouse

Castle Gate

Price River

**Martin
Yard**

Wildcat
elev. 47"

Load-out

Mohrland

Load-out
elev. 50"

Low horizon
board

Nolan Tunnel
elev. 56"

Helix
2.5 percent
grade, 24"
radius

Tunnel No. 1
elev. 46"

A

A

B   A

B

elev. 53"

Scale ¼"=12"
Grid=12"

LAYOUT AT A GLANCE

Overall size: 13.3 x 30 feet
Length of main line: 120 feet
  (excluding helix)
Locale: Wasatch Mountains, Utah
Layout style: walkaround oval
Layout height: 50"-56"
Benchwork: sectional open-grid
Track: Atlas code 83, No. 6 turnouts

Minimum radius: 30"
  main line, 24" helix
Maximum grade: 2.5 percent
Scenery: Carved extruded foam
Backdrop: .060" styrene; ¼"
  hardboard for low horizon
Control system: wireless DCC, main
  line signaled for CTC

Utah Railway, Martin Area

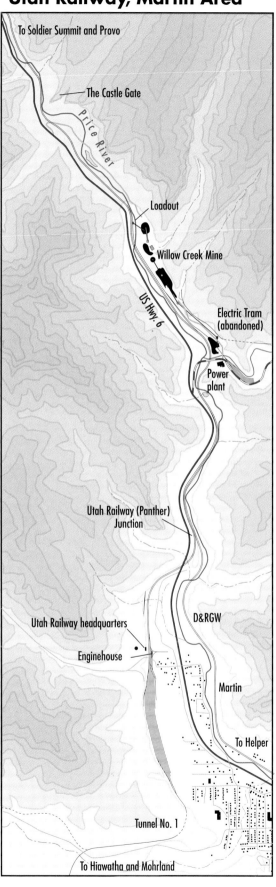

To Soldier Summit and Provo

The Castle Gate

Price River

Loadout

Willow Creek Mine

US Hwy. 6

Electric Tram
(abandoned)

Power
plant

Utah Railway (Panther)
Junction

D&RGW

Utah Railway headquarters

Enginehouse

Martin

To Helper

Tunnel No. 1

To Hiawatha and Mohrland

A Utah Railway coal train destined for the IPPX power plant in Lynndyl, Utah, approaches the Gilluly Loops. Two MK5000Cs lead, with a Genessee & Wyoming engine third in the consist. This view is from May 2004; the MK locomotives were repowered with EMD engines in 2002. *Mike Danneman*

Two EMD helpers push hard on the tail end of a manifest freight train as it enters Nolan Tunnel. *Bernard Kempinski*

tral corridor. The BNSF contracted with Utah Railway to operate its coal and local trains. BNSF engines are now frequent visitors to the line.

To handle the new business, Utah Railway leased an additional eight loco-motives, bringing its fleet to 21 engines. The UR also got access to new coal cus-tomers along the former D&RGW/SP lines including Savage Coal Terminal near Price and Cyprus Amax's Willow Creek mine near Castle Gate.

Also under the agreement, UP granted Utah Railway overhead traffic rights across the line between Utah Railway Junction and Grand Junction, Colo. In a further expansion, UR began operating BNSF trains into Ogden, Utah in Sep-tember 1997.

Over the years, UR's corporate owner-ship changed hands several times, mainly to mining concerns. In 2002, UR's then-parent company, Mueller Industries, sold the railroad to the Genesee & Wyoming, a short line and regional railroad holding company. UR still maintains a separate identity, but G&W engines began arriv-ing on the property soon thereafter.

Today's G&W's UR operates more than 423 miles of track between Grand Junction, Colo., and Provo, of which 45 miles are owned, and the rest operated under agreements with BNSF and Union Pacific. The company still hauls a signifi-cant amount of coal – just over two-thirds of the 90,000 carloads hauled each year.

While coal is the key factor in the operation, the transcontinental double-track line also saw heavy manifest freight traffic. During the Rio Grande era, it linked the Western Pacific with Denver. Now UP doesn't run much through freight on the line, but BNSF does. Pas-senger trains, in the form of Amtrak's *California Zephyr*, still make the scenic trip each way once per day.

Hauling freight and passengers over this mountain range has provided a chal-lenge to crews of both railroads for more than 100 years. Throughout this entire period, train crews changed at Martin or Helper at the eastern foot of the Wasatch Mountains. Over the years, 2-8-8-2s and 50-ton hoppers have given way to SD50s and aluminum 110-ton hoppers, but the task of shoving westbound coal up the canyon remains.

Z scale track plan

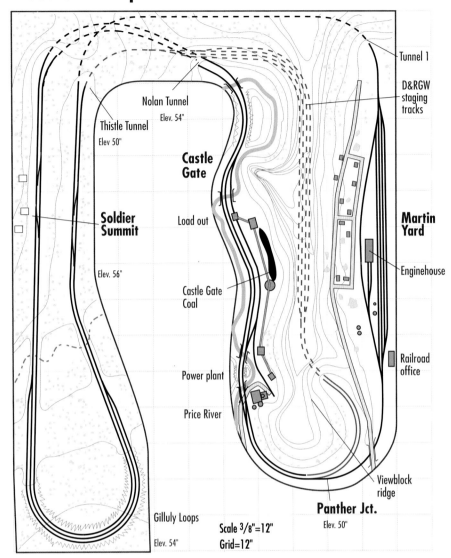

LAYOUT AT A GLANCE

Overall Size: 11 x 15 feet
Length of main line: 62 feet
Locale: Wasatch Mountains, Utah
Layout style: walkaround, double-track oval
Layout height: 50"-56"
Benchwork: sectional open-grid
Track: Micro-Trains code 55 flextrack

Turnouts: Micro-Trains 13D
Minimum radius: 19.3" main line, 15" D&RGW
Maximum grade: 2.5 percent
Scenery: Layers of extruded foam, carved
Backdrop: .060" styrene sheet
Control system: wireless DCC, main line signaled for CTC

There is surprising variety in the ter-rain as the railroad climbs to Soldier Summit. The arid, rocky terrain of the East Desert creates a bleak backdrop for the town of Martin. Cottonwoods and small shrubs, bright green in spring and summer, hug the Price River as it cuts the precipitous canyon followed by the rail-road and Highway 6. The topography opens up at summit. Evergreens, aspens, and fields of grass cover the saddle, creat-ing a verdant vista.

Z scale track plan

The smallest scale, Z, has made great strides in recent years. It is moving

Morrison-Knudsen locomotives ease their train under the coal load-out at Wildcat in this February 2004 scene. *Mike Danneman*

The Utah Railway's Martin office sits on a shelf above the engine terminal. Two EMD helpers await their next assignment. *Bernard Kempinski*

beyond the novelty, layout-in-a-briefcase stage to a point where one can consider designing and building a realistic prototype-based operating layout. Many North American prototype models are available, including modern diesels. High-quality Z scale freight cars are continually being released. Micro-Trains has introduced a range of realistic track and turnouts proportioned to North American specifications.

The layout features both slopes of the climb over the Wasatch. The UR gets most of the attention, while the Rio Grande or UP provides a parade of through trains to add variety and action. The time period of the layout is loosely based on the modern era, because there still isn't a complete selection of engines and other rolling stock to allow completely accurate prototype modeling in Z scale.

Schematically the plan is a simple double-track oval with separate staging yards for each of the two railroads. The UR Martin Yard is visible, while the D&RGW Yard is hidden under scenery. Since switching is not a strong point in Z scale, the main operational pattern is based on through and unit coal trains with helper action. The primary pattern for coal traffic is loads west, empties east. A double-track continuous-run plan is well suited for this type of operation, since trains that end their journeys are immediately restaged for the next run. By virtue of their enclosed cars, manifest freights can run in either direction.

The layout allows for grand scenic vistas that will please railfans. The track is deliberately simple, yet prototypically based to keep cost and maintenance at a minimum.

The plan starts at Martin and runs up the east slope to Soldier Summit. The western slope includes one of the Gilluly loops and ends at the Thistle Tunnels.

In the modern era, Martin was a crew-change point with limited yard switching. Thus the plan has a minimal yard, primarily to serve as a visible staging area. Sidings and staging tracks are long enough to handle trains of three units and 16-20 cars. The D&RGW yard is similar, but hidden behind scenery as the focus on this layout is the UR. You could easily reverse this arrangement if you wanted to highlight the Rio Grande.

Four locomotives lead a string of IPPX empties past the signature Castle Gate rock formations. The cliff on the left was once as imposing as the one on the right, until it was blasted to make room for a highway. *Ron Burkhardt*

A former Santa Fe SD45 and a Utah SD40 team up as pushers on this westbound Union Pacific manifest freight at Nolan Point. *Bernard Kempinski*

Several Utah Railway Alcos do an impressive imitation of a steam locomotive as they assemble a loaded coal train at Hiawatha in September 1981. The first, third, and fourth engines are ex-Santa Fe RSD-15s, and the second locomotive is a former Chesapeake & Ohio RSD-12. *William T. Morgan*

Even though the Utah Power plant was adjacent to the main line, it did not receive coal by rail. Coal for the plant is mined on site. *Bernard Kempinski*

Southern Pacific and Union Pacific engines lead empties past Willow Creek Mine just east of Castle Gate. The mine was dismantled in 2005. *Bernard Kempinski*

With just under 2.5 scale miles of visible main line, the plan can't capture the full nature of the prototype's 50-mile trip over the summit. However, by including several signature scenic vignettes as well as a 3 percent grade, the plan attempts to capture the drama and excitement of this mountain railroad.

Upon leaving Martin, the main line passes Panther (also called Utah Railway) Junction where the D&RGW enters the layout. The 3 percent grade begins at this point as the line follows the Price River. Willow Creek Mine hugs the view-blocking mountain and is the only on-line industry in the layout.

Next, the line passes spectacular Castle Gate. There is sufficient room in the plan to include a nearly scale-height representation of this famous rock formation.

The tracks curve into Nolan Tunnel, then emerge to run west past the abandoned homes on Soldier Summit. Though there is no siding here, helper engines can cut off and use the crossovers to return to Martin. Descending a 2 percent grade, trains head to Gilluly Loop, then into the Thistle tunnels.

On the model plan, the track in one tunnel connects to the UR Martin yard, while the other goes to D&RGW staging. The dispatcher has to align the trains to the correct track before they enter the tunnel.

A loaded train emerging from the west end of Tunnel 1 is automatically staged for a new run up the hill. Likewise, empties would depart Martin Yard for the notional coal mines, but would emerge at Thistle as another empty train returning from the power plants in Utah and Nevada. In effect, the tunnel represents a loads-in/empties-out type industry.

Although it's a small layout, it has ample room for spectacular scenery. Mountains, rock faces, and cliffs could be built up with layers of Styrofoam. Hard-carved Styrofoam and Durham's water putty will capture rock strata. Add multiple layers of dirt, crushed rock, and ground foam to complete the look.

Even though the layout has a desert theme, several hundred trees and bushes will need to be represented. Scrub brush can be made with tips of Supertrees, foliage clusters, and twine. Aspens can be

hand-made using dried gypsophila. Chenille and bottle brushes make good evergreens.

The layout, as in the prototype, is sparsely populated and includes only a few structures, but they're interesting ones. These include the Willow Creek Mine, with a car load-out, conveyors, washers, and water treatment; the town of Martin, with its engine facility; and several abandoned homes at Soldier Summit. Given the small number of structures in the layout, one could spend lots of time and effort on scratchbuilding and super-detailing each structure.

A hallmark of the Soldier Summit line is the double-track main line dispatched with bi-directional Centralized Traffic Control (CTC) for maximum capacity. To reflect this fact and provide the authentic flavor of the prototype, the layout would ideally incorporate an automatic CTC system. In spite of the relatively short mainline run, there are five crossovers in addition to the single-tracked UR line. They should provide an interesting and realistic operational challenge. Digital Command Control will

Leased Southern Pacific and Union Pacific engines bring a string of empties into the yard at Martin. *Bernard Kempinski*

simplify operations and make a signalling and dispatching system possible.

With its emphasis on through trains, the operating scheme for the layout focuses on keeping trains moving. The layout has room for three trains in visible staging and another four in the hidden D&RGW yard. A typical operating session would have a dispatcher control the staging, while one or two operators take trains out of staging and across the layout. Since the yard activity is minimal, the dispatcher will control the yard track assignments and call crews. Nearly all westbound trains used helpers, and using them on the layout will provide a challenge to the dispatcher.

The short mainline run can be made to appear longer if crews run their trains at prototypical speeds. Coal drags climb the summit at 12 mph. At this rate, a train would require 12 minutes to cross the layout. Throw in a stop to cut off helpers, and perhaps add a meet with another train, and a run could last up to 30 minutes. On the other hand, Amtrak's

California Zephyr will streak across the layout, taking five or six minutes to cross.

Thanks to the loop design, trains finish their run by returning to staging. Thus each train can run twice or more during a session, providing an hour or two of intense activity without swapping cars in staging. The hidden D&RGW staging would make this "fiddling" difficult.

Many of the locomotives one would expect to see at Soldier Summit are now available in Z scale, including the EMD GP7, GP35, SD40-2, SD45, and E8, Alco PA-1, and GE C44. You can even get a modern EMD FP59 passenger engine and Superliner cars.

Over the years the Utah Railway ran primarily Alco RSD-4s and -5s and EMD SD39s, SD40s, SD45s, and a few FP45s, along with some leased power (mainly Burlington Northern units), so some modeler's license will be required. The attractive modern UR paint scheme was actually designed by a model railroader who worked for the railroad. Unfortunately, decals are not available in

Z scale. Modelers wanting to paint UR engines will have to resort to custom decals. The UR often borrows Union Pacific engines. Thus for the diesel modeler there is prototypical justification to run just about any western road engine.

Because of the track plan's broad curves, the layout could be built in N scale in the same footprint with just a few changes, namely increasing the parallel track spacing and allowing for slightly longer turnouts.

Utah Railway in HO

The HO Scale layout provides a different take on the UR. By limiting the design to just a portion of the east slope, there is room to include some of the fascinating mine branches and their related switching activities. This orientation suits the nature of HO scale, where uncoupling and backing cuts of cars is easier and more practical than in smaller scales.

The overall room size for this layout, at 13.3 x 30 feet, is right at our self-imposed limit of 400 square feet. This

Three Alco RSD-4s and an RSD-5 (No. 306, the second engine) bring a string of empty hopper cars to mines near Helper in the late 1970s. The locomotives, acquired in 1952 and 1955, were the Utah Railway's first diesels, and they wore a paint scheme of white with a red stripe. *Bob Gottier*

room width is actually a little narrow given the central peninsula and its end turn. An extra foot of room width would help make the aisles more comfortable, but this is about the narrowest one can design a plan with a 30" minimum radius and still have aisles of a reasonable width. Since the traffic levels on the layout are not too great, and only a few operators are needed, the comparatively narrow aisles in this plan should not present serious problems.

Including the Utah Railway run from Martin to the Hiawatha and Mohrland presents the opportunity to model the spectacular Gordon's Creek Trestle as well as several coal load-outs and a small desert town, which is now largely a ghost town.

The operating scheme for this layout is similar to the Z scale plan. Essentially, the Utah moves loads west, while empties run east. The D&RGW is also included to provide through coal and freight action, but the main focus is on the UR and its trains that work the coal mines.

Although the plan for Martin is condensed, it is closer to the prototype arrangement than the Z scale plan. Like the Z scale plan, the wye is omitted, as it is not needed in modern-era operations.

The helix is carefully designed to allow multiple uses. The outside track is for D&RGW trains. This track splits from the helix after one turn and heads to the Rio Grande staging tracks. The inside track has a turnout just inside the portal. This creates a pair of long sidings to stage a coal train for the Mohrland load-out. The UR track in Tunnel 1 must dip down to clear the lowest track of the helix before climbing back to the load-outs at Wildcat and Hiawatha.

A typical session would have two mine runs made up at Martin in the morning. These trains would head to Wildcat and Hiawatha. The load-out at Mohrland utilizes helical staging to create a loads-in, empties-out operation. The load-out at Wildcat would have a loaded train visibly staged there. The empty mine run would arrive from Martin and exchange its

empties with the loaded cars already staged there. While not exactly prototypical, it provides a second mine location to switch, without the need to individually load or empty the cars.

A horizon skyboard backdrop provides limited access to the D&RGW staging yard. This yard has only four tracks. An Amtrak train and two Rio Grande freights could be staged here. Amtrak runs one train each way in the morning. If one of the tracks was kept empty, it could be used to run locomotives around the train, allowing these trains to run in each direction during a session. If a wider space is available, more staging tracks could be added to avoid this bit of fiddling.

Helper operation is a key part of this railroad. In this plan, the helpers will have to cut off before entering Nolan tunnel. This is not prototypical, but there isn't room to model more of the mainline climb. Since there's no easy way to add helpers to the D&RGW trains in staging, they should be given enough locomotives so that helpers are not needed.

CHAPTER 8

New York City's Bush Terminal

A New York Cross Harbor Railroad Alco switcher pulls a Santa Fe boxcar under the corner of a Bush Terminal building on Second Avenue. The crossing in the foreground is from an old streetcar line. *Tom Flagg*

In a city known for some of the most famous skyscrapers in the world, the New York City buildings that most fascinated me as a young boy were the warehouses and docks known as the Bush Terminal. Modeling the terminal and its accompanying railroad, which features tight curves, street running, and numerous spurs winding among tall structures, would be a challenging but fascinating project.

Fascinating structures

Growing up in southern Brooklyn during the 1950s and '60s, I had ample opportunity to observe the Bush Terminal, usually from the back seat of my father's car as we drove past on the elevated Gowanus Expressway. From this lofty position looking over the plains of brownstone apartment roofs, my gaze was fixed not on the distant glittering Manhattan skyscrapers, but on the stark white eight-story concrete warehouses boldly emblazoned with the Bush Terminal logo. The seemingly endless maze of block structures, connected by bridges and catwalks, with shadowy alleys, crisscrossed by railroad tracks and adjacent to long-fingered piers captured my attention. I often wondered what went on in there.

That question went largely unanswered until nearly 40 years later. I learned that the Historical American Engineering Record (HAER) had documented the Bush Terminal. It seems the piers' decaying timbers created a drift hazard to marine navigation, so the wooden pilings had to be removed. Because of the complex's historical significance, the HAER documented the terminal and archived the results.

Using this information, I designed a layout for the 2003 issue of *Model Railroad Planning* that would fit on two small shelves. This highly condensed design omitted many of the prototype's most interesting features. This chapter features expanded designs that better capture the atmosphere and operation of the Bush Terminal Railroad.

Car float operations

The railroad and warehouse complex dates to the 1890s, when private developer Irving Bush decided to build a terminal on waterfront land in an undeveloped part of Brooklyn. He started construction in 1902. By World War I, the Bush Terminal was the largest multi-tenant industrial property in the United States.

According to the HAER records, "Bush Terminal was the first American example of a completely integrated manufacturing and warehousing facility, served by both water and rail, under unified management. Largely intact today, it remains the largest unified non-railroad terminal ever built in the Port of New

NYCHRR and freight railroad connections, circa 2001 (post-Conrail breakup)

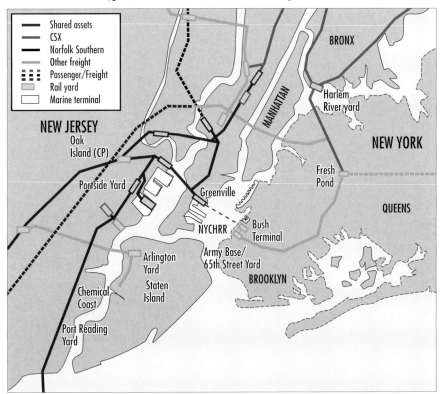

A New York Dock GE switcher drags a cut of cars from the car float lead. The subway cars are returning from shop work in New Jersey. *Gerry Landau; Bernard Kempinski collection*

York, and retains a rare survival of an isolated freight railroad served only by float bridge."

There were three distinct aspects to the 200-acre terminal, all operated as a single entity: manufacturing, warehousing, and transportation.

First were the seven quarter-mile long piers and more than 100 warehouses where break-bulk cargo was loaded to and from

In its heyday, Bush Terminal car floats could be found up and down New York Harbor. Here a tug is controlling two floats at one time. Note the blocking to secure the cars, the small crew house with vents, and the track arrangement. *Tom Flagg collection*

The Bush Terminal's car float transfer bridges are a pontoon design. The turnout points are on the bridge (visible on the track at left), with the frogs on the float. *Tom Flagg*

ships. Prior to the Bush Terminal, loading ships in New York harbor was a disorganized and inefficient process. Bush and his architects carefully studied the problem, and by adopting a series of simple, comprehensive and low-cost innovations they were able to offer shippers and customers highly efficient stevedore services. For example, Bush's piers were solidly built on fill so they could support railroad tracks. The pier sheds had battered (sloped) sides so that ships could use their onboard cranes without interference. The sheds also had a continuous set of overhead doors to make movement in and out of the sheds easier.

The second aspect to the terminal was the business of renting the large loft warehouses that so vividly captured my attention. Made from reinforced concrete, a relatively new construction material at the time, these cavernous buildings offered seemingly endless loft space to clients. The concrete construction allowed walls with large windows. They were all designed for excellent access to rail sidings. Heavy-duty freight elevators served each floor. Once a customer placed goods on the freight elevator, the Bush Terminal Corporation took over, transporting the goods via its own railroad and connections to the destination under one freight rate. This allowed shippers to avoid additional charges for drayage to and from railroad freight houses.

The terminal had its own power plant, trolley service, fire protection system, banks, and restaurants. It also offered a marketing service, including showrooms in Manhattan, which allowed companies to have a presence in the city without having to maintain a separate branch.

The Bush Terminal Railroad connected all of these disparate components. It provided rail links to other railroads via a land connection to the Long Island Rail Road and with transfer bridges and car floats to other railroads serving the harbor. The Bush Terminal operated several steam locomotives, two wooden steam tugboats, six steel three-track car floats (each with an 18-car capacity), and two wooden floats. The tugs and floats would pick up cars from around the harbor and bring them to the Bush Terminal transfer bridges. Freight cars would be moved to the First Avenue yards (which had a capacity of 2,000 cars) via the transfer bridges to be sorted and then delivered to piers, warehouses, or factories. In later years, the railroad ran several second-hand diesels including Also S-1s, an RS-3, and several GE 44-tonners.

Through the 1960s, containers became increasingly popular, and the break-bulk freight business declined. This eliminated Bush's shipping tenants, although some industries remained in the loft warehouses. The Bush Terminal Co. went out of business in 1972.

At that time the city of New York contracted with the New York Dock Railroad to take control of the facility and operate it. The New York Dock was an old company, founded in 1901, and had operated in three main locations in Brooklyn. In 1983, the New York Dock

Bush Terminal area, circa 1942

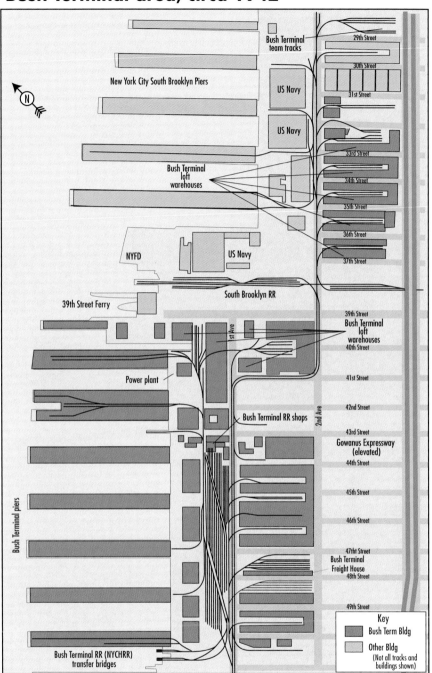

transferred ownership to the New York Cross Harbor Railroad.

By 1988, the upland business remained substantially intact, operated by the Industry City at Bush Terminal, but the waterfront piers had become hazards, having suffered from neglect and several fires. The Army Corps of Engineers removed the piers from 1978 to 1980. By 2007, many of the original Bush terminal structures had been renovated, redeveloped for alternative uses by high-tech industries and others.

Amazingly, the railroad car floats are still in operation under the stewardship of the New York Cross Harbor Railroad. The NYCHRR runs all three remaining railroad facilities along the Brooklyn Waterfront, including Bush Terminal, using three Alco S-1s, an Alco S-4, and two EMD NW2s. Its core business is moving corn syrup, lumber, rice, newsprint, electronics, plastic pellets, and bricks. The railroad also hosts intermodal traffic, primarily containers of refuse and scrap on flat cars.

The loft warehouses create an urban canyon for the tight curved sidings between buildings. Note the styles of the trackside loading docks. *Historic American Engineering Record*

Bush Terminal N scale plan, present day

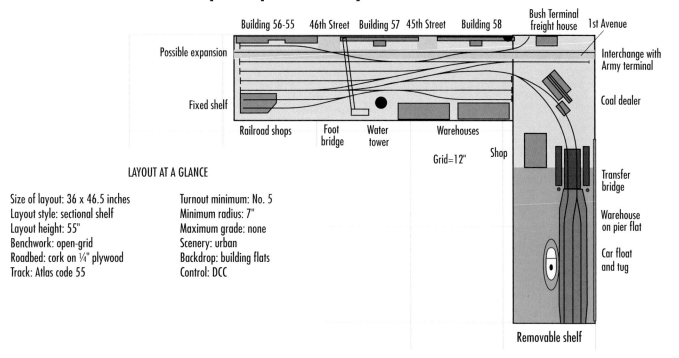

LAYOUT AT A GLANCE

Size of layout: 36 x 46.5 inches
Layout style: sectional shelf
Layout height: 55"
Benchwork: open-grid
Roadbed: cork on ¼" plywood
Track: Atlas code 55

Turnout minimum: No. 5
Minimum radius: 7"
Maximum grade: none
Scenery: urban
Backdrop: building flats
Control: DCC

On an average day the railroad moves 18 cars on its floats across the harbor to the former Conrail facility at Greenville, N.J. In addition to freight cars moving to the warehouses, the floats also transport New York City subway cars to maintenance facilities. In the First Avenue yard, the NYCHRR keeps its enginehouse, a bulk facility for transferring plastic pellets from freight cars to trucks, and a small intermodal terminal.

The NYCHRR interchanges with the New York & Atlantic Railway in Brooklyn for cars whose destinations are Long Island, the northeast United States or Canada. NYCHRR also connects with the New York City Transit subsidiary South Brooklyn Railway.

N scale plan

The N scale plan distills the real 200-acre facility down to a layout that can fit on two bookcase shelves. The layout requires a good deal of selective compression, even by N scale standards. Nonetheless, two 10 x 36-inch shelves provide just enough room for a convincing representation of the Bush Terminal plus some features from the surrounding area.

The shelf containing the transfer bridges is removable and can be stored separately. There are no turnouts spanning the joint between the shelves.

Bush Terminal tracks ran down Second Avenue, with spurs ducking between the warehouses at right. *Historic American Engineering Record*

HO scale track plan, circa 1950

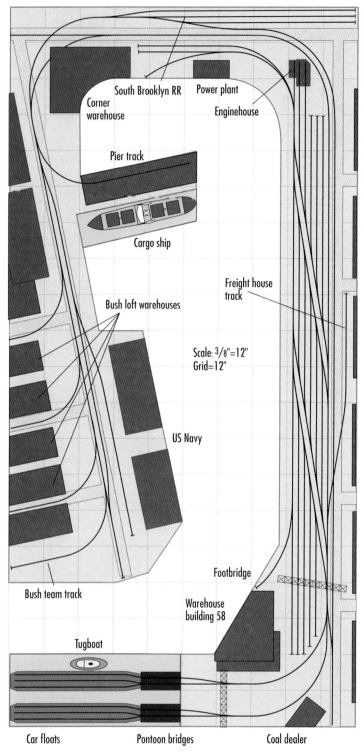

South Brooklyn RR

Power plant

Corner warehouse

Enginehouse

Pier track

Cargo ship

Freight house track

Bush loft warehouses

Scale: 3/8"=12"
Grid=12"

US Navy

Footbridge

Bush team track

Warehouse building 58

Tugboat

Car floats

Pontoon bridges

Coal dealer

LAYOUT AT A GLANCE

Overall size: 10 x 20 feet
Length of main line: 50 feet
Period: 1952-53
Layout style: point-to-point
Layout height: 52"
Benchwork: sectional open-grid
Roadbed: cork on ½" plywood, cookie-cutter style

Track: Atlas code 83
Turnouts: Atlas No. 6, handlaid curved turnouts
Minimum radius: 24" main, 18" spurs
Maximum grade: None
Backdrop: .060" sheet styrene
Control system: wireless DCC, hand-thrown turnouts

The basic arrangement of the tracks and transfer bridge on the plan follows the prototype. Of course, the number of tracks is heavily compressed.

The small plan focuses on the modern period with the NYCHRR working the terminal. The scaled-down operations of the modern period suit this small railroad. This also permits using some of the fine-running N scale switchers such as the Arnold S-2 or Life-Like SW1200. The sight of Alco switchers operating in streets with modern cars, twisting around warehouses and onto float bridges, is a unique spectacle worth modeling.

The limited depth of the shelf leaves insufficient room for tracks running per-pendicular to the waterfront between the loft warehouses. Connecting the shallow loft structures with overhead bridges, although a more common feature in the adjacent Army supply base than Bush Terminal, is used here to hide the back-drop and create the illusion that the buildings extend deep into the scene.

I'd suggest using a Digital Command Control system. That would provide the flexibility of allowing two locomotives and crews to work the layout at a time, although one crew would work fine too.

Operating this layout would be similar to the process used by the prototype as described earlier. Unlike the prototype, we have the luxury of easily moving an ille-gally parked car blocking our way, instead of having to find the owner or wait for it to be towed.

I'd suggest using the car floats for staging without making them removable. An operating session would start with inbound cars waiting on the floats. To extend the session, one could fiddle the cars on the float to provide variety. The floats are nearly scale length and hold eight 50-foot cars. Remember to include a locomotive on the float when it is ready to "depart," as it will be needed when the float reaches Greenville Yard. Remember that the car float should not be used as a classification track: You should pull cars from the float in groups and use the First Avenue Yard to sort cars.

There are three "industries" or loca-tions where cars can be spotted: Loft Warehouse A, the warehouses near the front edge, and the interchange track. You can make switching more interesting by

This aerial view shows the extent of Bush Terminal and its relation to New York City. Manhattan is visible through the haze at top center. *Tom Flagg collection*

Alternate HO plan with two piers

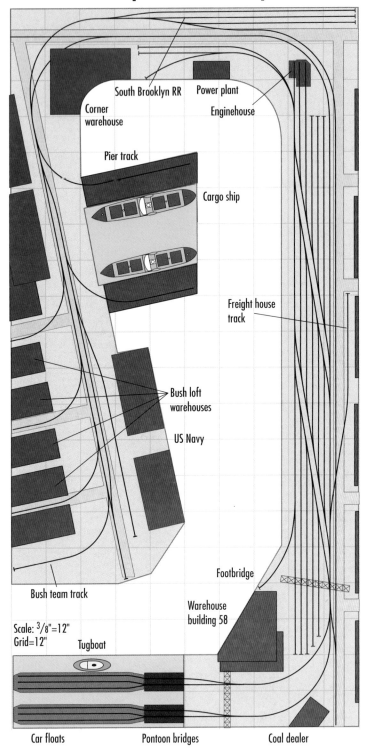

South Brooklyn RR Power plant

Corner warehouse Enginehouse

Pier track

Cargo ship

Freight house track

Bush loft warehouses

US Navy

Footbridge

Bush team track

Warehouse building 58

Scale: $^3/8$"=12"
Grid=12"

Tugboat

Car floats Pontoon bridges Coal dealer

designating specific spots on each track for certain customers. First Avenue Yard has two ladders and seven tracks that can be used for classification, car storage, and layover. Some of these yard tracks can also be designated for plastic pellet transfer to trucks and for intermodal traffic in the modern period. The shops can be used for bad-order cars and servicing engines.

The HO plan

The HO plan is designed for 10 x 20 feet. This plan is still selectively compressed, but it more accurately simulates the yard and industries it served. As in the N scale plan, the heart of the layout is First Avenue Yard. Although surrounded by buildings and piers in the prototype, the yard is situated on the edge of the aisle. The distinctive buildings between the yard and the aisle have been omitted to improve access for operations.

The wall behind the yard is lined with shallow building flats. These buildings are generally one or two stories. The Bush Terminal freight house track is parallel to the yard, instead of perpendicular as on the prototype. There wasn't room to include it in its proper orientation, but omitting it would remove an important destination for cars.

This plan allows room to do a decent job of simulating the loft buildings, including some of the tracks that run perpendicular to the waterfront. Adding the Navy warehouses to the aisle side of the tracks, opposite the loft warehouses, creates the effect of an urban canyon. These structures can be made removable for access for maintenance and construction.

The plan includes one track extending out to a pier, running inside the pier shed. The shed should include the battered sides adopted by Bush. The plan includes only one pier, as adding others would create access problems for the loft warehouses.

By using remote switch machines and uncoupling magnets it might be possible to add a second pier as shown in the drawing. Accepting the potential access problems would allow including a second pier track as shown in the drawing. Several cargo ship models are available in 1/96th scale that could be used here.

The HO plan includes the interchange with SBK. Among other traffic, Bush Terminal interchanged subway cars here. They were floated across the harbor for maintenance.

The plan includes two car floats, as on the prototype. Walthers released HO kits for car floats and transfer bridges several years ago. Some modifications will be necessary, as the car floats used by New York railroads were somewhat atypical. The turnout frog for two of the three tracks is located on the float, while the points are on the transfer bridge. The tracks taper on the transfer end.

The floats came in different lengths. The plan includes two 320-foot floats,

First Avenue Yard is fairly empty in this 1930s view. The warehouses have external cranes to lift cargo to upper floors – a common arrangement before interior freight elevators became common. *Tom Flagg collection*

which should allow staging of 16 to 21 40- to 60-foot cars per float. This is enough to create an interesting operation session for two or three operators.

The plan uses all conventional track components, except for one curved turnout. Using open-grid construction, the layout should be relatively easy to build and get in operation quickly. You can then take your time scratchbuilding the structures needed to give the layout its distinctive appeal.

CHAPTER 9

[**1865**]

U.S. Military Railroad
City Point and Army Line

The proud crew members of wood-burning 4-4-0 No. 133 pose with their charge in front of the blockhouse at City Point. *U.S. National Archive*

Building a layout based in the 1800s can be a challenge, but it can also be rewarding from both historical and modeling standpoints. The Civil War is well documented, with more photo coverage than some 1900s-era short lines. The story of the Civil War-era railroads in and around City Point lends itself nicely to a medium-sized layout.

City Point history

By June of 1864, General Ulysses S. Grant had assumed command of Union ground forces. Grant's grueling, bloody spring campaign in northern Virginia led him to believe that heavily fortified Richmond, the Confederate capital, could not be taken by a direct assault. Instead, Grant decided to maneuver the Union Army of the Potomac 20 miles south of Richmond with the intent of taking Petersburg, a vital railroad hub that linked Richmond with the rest of the southern states. If Petersburg fell and its rail lines were cut, Richmond would be nearly isolated and untenable.

Union engineers demonstrated their overwhelming logistical might by building a mile-long pontoon bridge across the James River to reach Petersburg. Although the surprise river crossing was successful, the Union army failed to take Petersburg. Realizing that further frontal assaults against Petersburg's growing entrenchments would result in horrific casualties, Grant ordered a siege. The Union army settled in for a protracted investment that the embattled Confederates were unable to lift.

The Union's rail and naval superiority would play a key role in the siege. Grant understood the importance of supplying his army during the siege. So on June 13, as both armies dug in, he ordered his chief quartermaster general, Brevet Major General Ingalls, and commander of the U.S. Military Railroad, General McCallum, to build a harbor at City Point, Va., and a rail line from there to points behind his army outside of Petersburg.

The existing City Point harbor was in poor condition, and the rail line connecting it to Petersburg was either torn up or the wrong gauge and in a state of extensive disrepair. Under Ingalls' and McCallum's control, the City Point harbor and railroad depot were operational within 30 days. The original order for the depot at City Point called for temporary facilities, and they were built in that manner. However, as the siege continued, the USMRR construction crews rebuilt the temporary wharves and expanded the facilities. It grew into a large but extremely efficient supply depot and intermodal transportation center. By the end of the campaign, it had grown to more than 280 buildings,

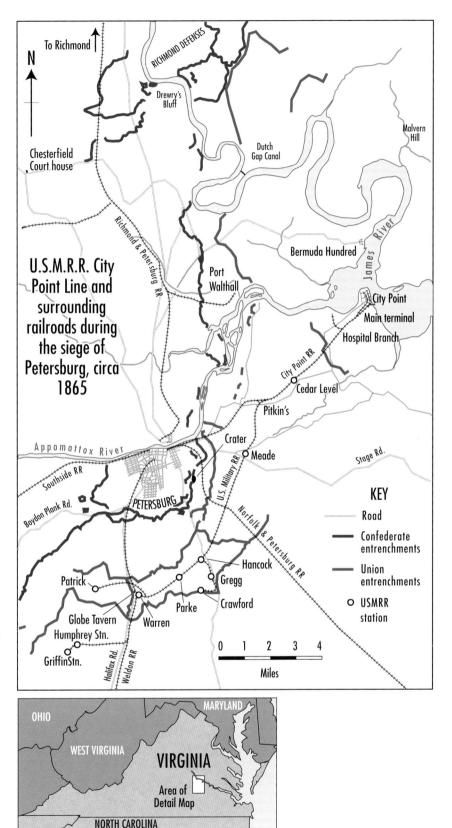

eight wharves covering eight acres, and warehouses with a capacity exceeding 100,000 square feet.

We are fortunate that the chief engineer and general superintendent of the

USMRR commissioned William Merrick to survey and create detailed maps of the City Point terminal and rail lines. The statistics and prototype maps in this chapter are based on these maps, which

Civilians as well as soldiers camped at City Point. The conical tents, quite common at the time, were called Sibley tents after a Union general. *U.S. National Archive*

The sawmill at the wharf has been busy, judging by the lumber piles and loaded cars. Caissons lined up at the wharf are ready to be moved out. *U.S. National Archive*

are available at the U.S. National Archives.

A network of 21.6 miles of track, including 1.2 miles on trestles at an average height of 21 feet, extended from the wharves to locations behind the Union lines. The photo on page 94 shows a typical trestle. Railroad facilities at City Point included a nine-locomotive enginehouse, a turntable, a car ferry ramp, repair shops, fuel platform, water tanks, dispatcher's office, and the railroad headquarters.

Two three-track yards and several spurs provided ample capacity for offloading cargo from ships to freight cars or wagons. At any given time, 150 to 200 ships were anchored off City Point waiting to unload cargo. More than 390 ships routinely worked among City Point and other Union ports and supply centers, making City Point one of the busiest harbors in the United States.

The wharves extended eastward more than a half mile along the southern banks of the James River. The wharf facilities covered more than eight acres, with a quarter of that under cover.

An ammunition ship explosion on August 9, 1864, the work of a Confederate saboteur, destroyed a good portion of the wharf. Construction crews quickly repaired the damage while also building a dedicated ammunition wharf a half-mile down the river to isolate the munitions from the rest of the harbor. The new ammunition wharf extended 500 feet into the river, with a rail line for unloading ships. Attached to this wharf was a dock for storing coal for ships. (The USMRR locomotives at that time burned wood instead of coal.)

The rest of the harbor was divided into specific wharves for different commodities. Major departments, such as the repair shop, managed their own wharves. Forage for animals was another major commodity, and that department had its own warehouses and wharf.

The Army also shipped freight cars on car floats from Alexandria to City Point. Photographs of the area show a single-track car ferry ramp. Since car ferries at the time had eight tracks, they would have to be moved to unload each track. This was rather inefficient, and it's unclear how much cargo was moved this way. The ferries' primary purpose was

West end of City Point line

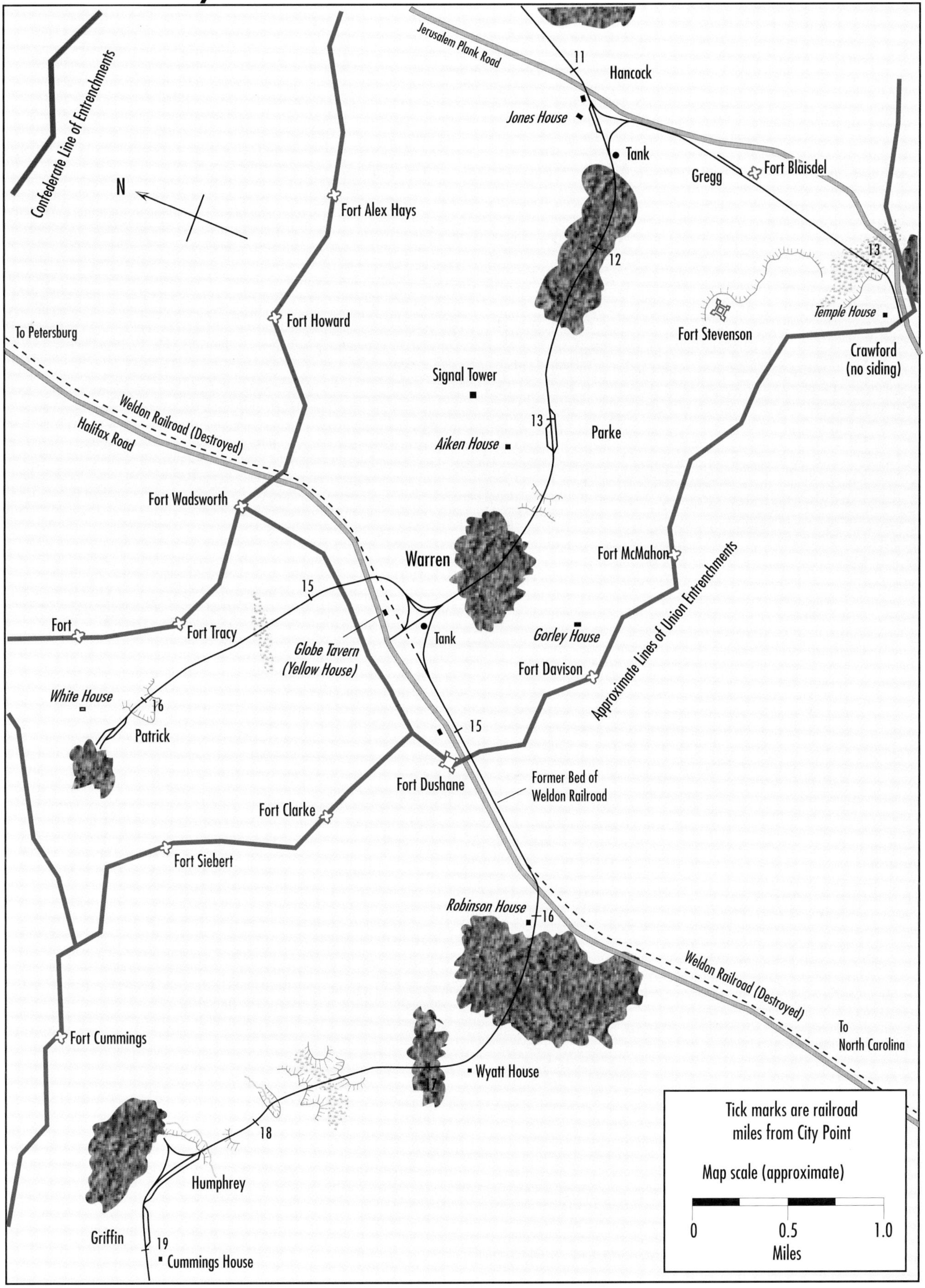

Jerusalem Plank Road

11

Hancock

Jones House

Tank

Gregg

Fort Blaisdel

13

Temple House

Crawford
(no siding)

Confederate Line of Entrenchments

N

Fort Alex Hays

12

Fort Stevenson

Fort Howard

To Petersburg

Signal Tower

13

Parke

Weldon Railroad (Destroyed)

Halifax Road

Aiken House

Fort Wadsworth

Warren

Fort McMahon

Approximate Lines of Union Entrenchments

15

Fort

Fort Tracy

Tank

Gorley House

Globe Tavern
(Yellow House)

Fort Davison

White House

16

Patrick

15

Fort Dushane

Former Bed of
Weldon Railroad

Fort Clarke

Fort Siebert

Robinson House

16

Weldon Railroad (Destroyed)

To
North Carolina

Fort Cummings

Wyatt House

17

18

Humphrey

Griffin

19

Cummings House

Tick marks are railroad
miles from City Point

Map scale (approximate)

0   0.5   1.0
Miles

A car ferry holding several boxcars (left) waits at the ferry ramp at City Point. Several ships are anchored in the harbor awaiting unloading. *U.S. National Archive*

probably to bring fresh cars and locomotives to City Point.

On the main line, the USMRR built sidings, platforms, water tanks, trestles, wyes, and a telegraph line for dispatching. The railroad was built with little grading, resulting in steep grades, and with ties that were basically logs which still had their bark. However, the system effectively linked all areas of the depot at City Point with the ever-expanding front around Petersburg. One observer noted that a train moving across the line in the distance looked "like a fly crawling over a corrugated washboard."

The rail lines greatly reduced the time required to get supplies and soldiers to the front. To soldiers accustomed to eating hardtack and salted bacon, the fresh meat and loaves of fresh bread now delivered daily was a huge morale boost. The operation was so efficient that the bread could be delivered while still warm to soldiers in their trenches.

By contrast, Southern soldiers were slowly starving, because the worn-out rail system was unable to bring in the plentiful food from the Shenandoah Valley and other secure agricultural areas to Richmond.

Operations and facilities

By April 1865, the line included two branches and 11 stations, each named after the commanders of the units they served. The stations were usually simple affairs with either a stub or passing siding serving loading platforms (see page 96). In some cases, a station house was present. An unnamed station at the west end of City Point also was the site of a wooden blockhouse with a central cupola replete with firing slits and observation ports.

At its peak, the line saw 15 trains daily each way plus specials, hauling troops and a daily average of 1,400 tons of supplies from City Point to the front lines. On return trips, the trains carried sick and wounded soldiers back to the hospitals at City Point. Several special hospital cars were kept in almost constant use.

Two passenger trains ran each way daily for mail, officers, and others going to and from the front. The line moved an average of 63,000 passengers per month during the last three months of the siege.

The number of locomotives and freight cars in service on the City Point and Army line isn't known. National Archive photos show at least 11 locomotives, and toward the end of the war, the line requested 24 additional engines and 275 boxcars, but they arrived after the siege had ended.

Repair shops, barracks, tent campgrounds, bakeries, stables, slaughter houses, warehouses, seven hospitals, laundries, sutlers (private vendors selling non-military supplies), and volunteer civilian commissions were present at City Point. Even specialized workers and agencies such as secret service personnel, scouts, and photographers had separate buildings for their own use.

General Grant established his headquarters on the bluff overlooking the harbor. He refused to reside in the elaborate Appomattox House, where General Ingalls set up camp. Instead, he lived in a simple log cabin where his wife and son visited.

U.S.M.R.R. City point terminal and logistics base

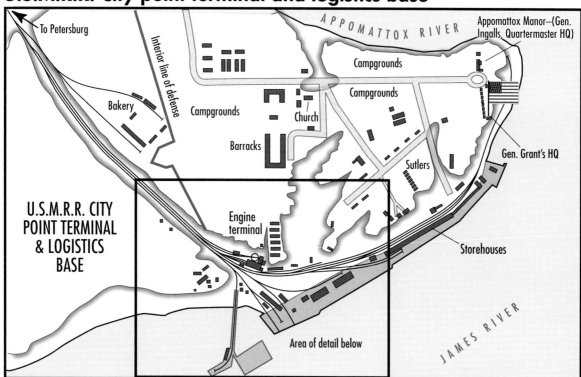

U.S.M.R.R. Engine terminal

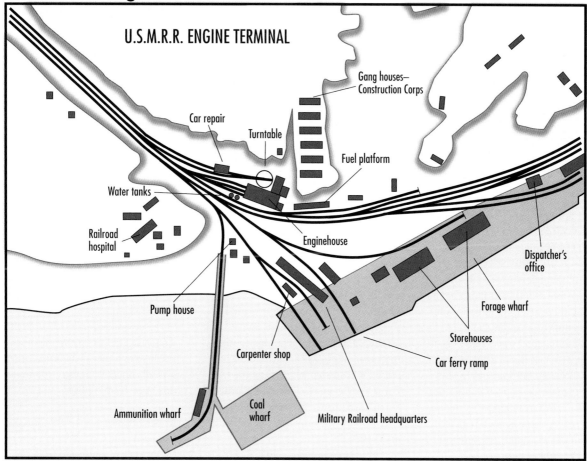

Source: Military Railroad map of City Point, Va., principal terminus of City Point and Army Railroad line and base of armies operating against Richmond. Drawn at Office of Chief Engineer and General Superintendent, Military Railroads of Virginia, Alexandria, Va., June 1865, from actual survey by Wm. M. Merrick, engineer and draftsman.

Many trestles were needed to cross creeks and other low areas. This lightweight trestle is typical of those found on the line. *U.S. National Archive*

A passenger car was reportedly lined with lead so that dignitaries could visit the front lines in relative safety. On March 23, 1865, President Lincoln did just that. He, Grant, and several other officers rode the car to observe the fighting. Lincoln remained at City Point, staying on a riverboat until April 3, when he visited the ruins of captured Richmond.

The depot commander controlled the overall operation and water traffic. Individual department chiefs managed their areas as necessary, but General Ingalls dictated that the depot commander submit a daily transaction report listing all receipts, issues, and balances. Ingalls' strict control greatly assisted in accountability and ensured proper prioritization of the supplies. The system was organized so that requests from the field for supplies could be filled within 24 hours.

A railroad dispatcher operated from a two-story building on a wharf. From there he controlled operations on the railroad using the telegraph. Earlier in the war, General Haupt had instituted time-table-and-train-order operation for the USMRR and formalized the system's operation procedures. Prior to this, local commanders frequently confiscated freight cars as storage sheds and generally interfered with rail operations to suit their own needs. Although Haupt had left federal service by the time the City Point line began operations, the procedures he instituted remained. Thus the City Point and Army line ran efficiently on a schedule.

Ten sidings along the line allowed meets between trains, and wyes at Pitkins, Hancock, Warren, and Griffin stations allowed turning engines.

War's end

The City Point line performed its job well, but on April 3 the Union Army abandoned the railroad when Union forces took possession of Petersburg. Lee's army retreated to Appomattox, where it surrendered. The physical components from the City Point and Army line, such as the rail, were used to rebuild portions of the devastated Confederate railroads. The structures and other facilities at City Point were either dismantled or sold to private concerns. Today little remains of this once-bustling base.

Modeling the Civil War railroads

Modeling the Civil War era is attractive for several reasons aside from the fascinating history. One can fit a lot of railroad in a small area, because the railroad equipment of the period was relatively

U.S. Military RR, HO scale, circa 1865

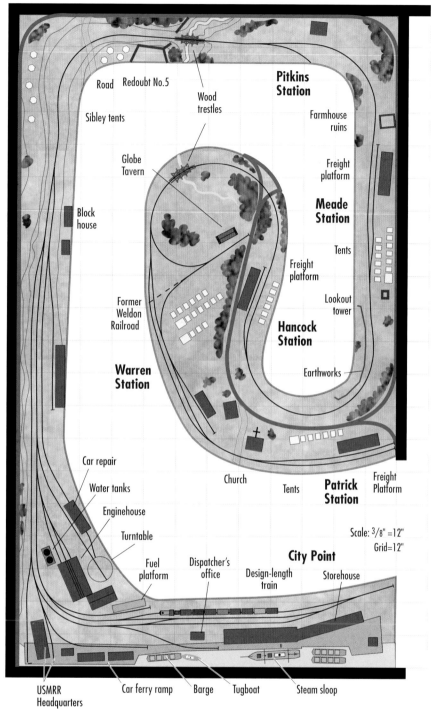

Road • Redoubt No.5 • Wood trestles • **Pitkins Station** • Sibley tents • Farmhouse ruins • Globe Tavern • Freight platform • **Meade Station** • Block house • Tents • Freight platform • Former Weldon Railroad • Lookout tower • **Hancock Station** • **Warren Station** • Earthworks • Car repair • Church • Tents • **Patrick Station** • Freight Platform • Water tanks • Enginehouse • Turntable • Scale: 3/8" =12" • Grid=12" • Fuel platform • Dispatcher's office • **City Point** • Design-length train • Storehouse • USMRR Headquarters • Car ferry ramp • Barge • Tugboat • Steam sloop

LAYOUT AT A GLANCE

Overall Size: 12 x 20 feet
Length of main line: 73 feet
Prototype: USMRR
Locale: Petersburg, Va.
Layout style: walkaround point-to-point
Layout height: 50"
Benchwork: sectional open-grid

Roadbed: cork on ½" plywood, cookie-cutter style
Track: code 83
Turnouts: No. 5, handlaid
Minimum radius: 24", 15" in wye
Maximum grade: none
Backdrop: .060" styrene sheet
Control system: wireless DCC, hand-thrown turnouts

Three locomotives stand ready for service. Note the tie placement on the foreground track. *U.S. National Archive*

Most stations were temporary, crude affairs. At Cedar Level station, supplies were stacked on a car-level loading platform at the siding. *U.S. National Archive*

small. Most boxcars were only 28 feet long, and the era's common 4-4-0 American steam locomotives weren't much bigger. Train lengths were modest, typically 15 to 16 freight cars. Such a train in HO scale is only about five feet long. Because the equipment is small, tighter curves can be used.

Thanks to pioneering photographers such as Matthew Brady and Major Andrew Russell, there are many excellent photographs of the Civil War era, including railroad subjects. City Point in particular was well documented by photographers. Additional National Archive photos continue to be placed online, making research even easier.

Official records from both sides of the conflict, along with several other sources, provide detailed descriptions of railroad operations. Due to heavy demands of military operations, Civil War railroads were very busy, yet suitable for timetable-and-train-order (TTO) operation. This dispatching method was in its infancy in this era, and researching the nuances to duplicate TTO operations in the proper manner for the period would be a rewarding pastime once the layout was built.

In selecting a scale, modelers may be limited by the availability of equipment. Engines were predominantly 4-4-0s with 4-6-0s and 0-8-0 Camelbacks used by a few railroads. While N scale has several suitable freight cars on the market, there is only one 4-4-0 readily available, and small steam locomotives are often not reliable in N scale.

Modelers in HO have the broadest selection of appropriate equipment. The old Mantua and Rivarossi 4-4-0 models are correct for the era, but they usually need some work to run reliably. Some of the more-recent, fine-running 4-4-0 models depicting later prototypes could be backdated to the Civil War era by re-detailing them. Several manufacturers, including BTS and Alkem Scale Models, offer appropriate kits for freight cars with the correct trucks and details.

Fine-scale 4-4-0s are scarce in O scale, but some manufacturers, such as SMR Trains, have made beautiful models in limited runs albeit at relatively high cost. The O scale track plan exploits the availability of these engines. A layout in G scale (1:32) is an interesting possibility as

U.S. Military RR, HO scale alternate plan

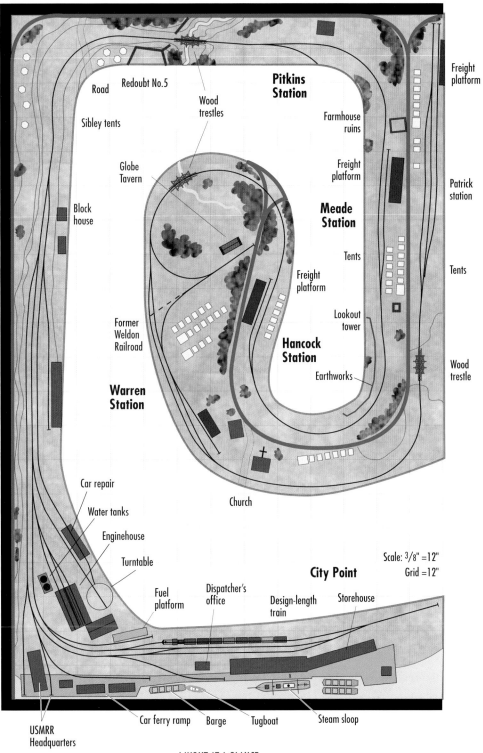

Freight platform

Road

Redoubt No.5

Wood trestles

Pitkins Station

Sibley tents

Farmhouse ruins

Patrick station

Globe Tavern

Freight platform

Meade Station

Block house

Tents

Tents

Freight platform

Former Weldon Railroad

Wood trestle

Hancock Station

Earthworks

Warren Station

Church

Car repair

Water tanks

Scale: 3/8" =12"

Grid =12"

Enginehouse

Turntable

City Point

Fuel platform

Dispatcher's office

Design-length train

Storehouse

USMRR Headquarters

Car ferry ramp

Barge

Tugboat

Steam sloop

LAYOUT AT A GLANCE

Overall Size: 13 x 20 feet
Length of main line: 88 feet
Prototype: USMRR
Locale: Petersburg, Va.
Layout style: walkaround point-to-point
Layout height: 50"
Benchwork: sectional open-grid

Roadbed: cork on ½" plywood, cookie-cutter style
Track: code 83
Turnouts: No. 5, handlaid
Minimum radius: 24", 15" in wye
Maximum grade: none
Backdrop: .060" styrene mountain horizon
Control system: wireless DCC, hand-thrown turnouts

Tents could be found almost anywhere at City Point. This view shows typical boxcars of the era and the basic construction of structures. *U.S. National Archive*

there are hundreds of military figures and supporting details available. However, standard gauge 4-4-0s and appropriate cars have not been offered.

A couple of challenges exist in modeling this era in any scale. First, railroads used stub switches, which would have to be handlaid. Also, cars used link-and-pin couplers, which will be nearly impossible to model.

Fortunately, several manufacturers make appropriate detail parts. Musket Miniatures and K&M offer HO and N scale figures, military equipment, and scenery items appropriate for the era. Narrow gauge manufacturers also offer many components that can be readily adapted to the Civil War period.

HO layout design

The basic point-to-point layout fits in a 12 x 20-foot room. The plan assumes walls on all sides, but entry access is needed at lower right. There is no off-layout

staging as the line is self-contained. The track plan follows the prototype arrangement quite closely with a few stations and wyes omitted and the length compressed. The wharf area includes nearly all the prototype's tracks, which evolved significantly during the campaign. The track plans are based on the arrangement shown in Merrick's survey.

Over the route, the prototype line climbed a couple hundred feet as it left the river. The plan should include a gradual (0.5 percent) grade to Warren Station, which will be 2" or 3" higher than City Point. However, short sections of steep grades in and out of ravines and over hillocks will represent the temporary nature of the grading. The wooden trestles should be modeled with materials simulating unfinished lumber or tree trunks.

Flextrack would simplify construction, but the most prototypical look will come from handlaying track on rough-hewn ties. Special care must be taken because

model trains are more sensitive to irregularities in track than the sure-footed prototype 4-4-0s.

The wye at Warren station is an important location on the plan, as it's the only place out on the line to turn locomotives. The actual orientation of the wye is slightly changed from the prototype to better fit the available space. The double-ended siding here will enhance operation on the layout, although it's in a slightly different location compared to the real thing.

The Hancock and Meade stations are simple stub-ended sidings. These could be made into double-ended sidings to better reflect the prototype at the cost of extra complexity.

If room is available, you can extend the line behind Meade Station as shown in the optional plan. Placing Patrick Station in that location would increase the length of run from Warren's station, but does not change the overall operation. The expan-

U.S. Military RR, O scale

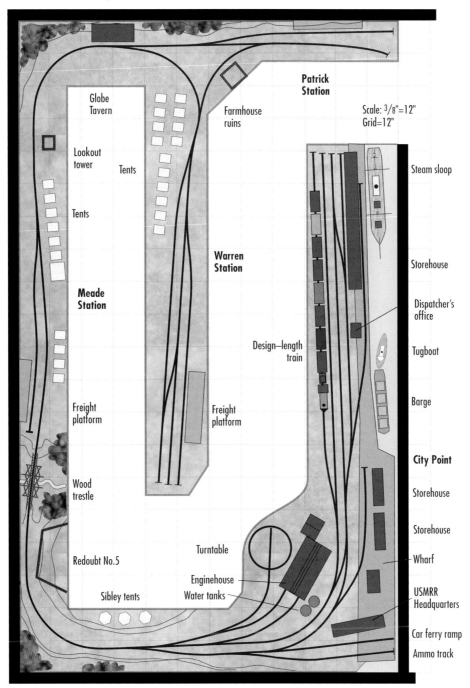

Globe Tavern

Lookout tower

Tents

Tents

Meade Station

Freight platform

Wood trestle

Redoubt No.5

Sibley tents

Farmhouse ruins

Patrick Station

Scale: 3/8"=12"
Grid=12"

Warren Station

Design—length train

Freight platform

Turntable

Enginehouse

Water tanks

Steam sloop

Storehouse

Dispatcher's office

Tugboat

Barge

City Point

Storehouse

Storehouse

Wharf

USMRR Headquarters

Car ferry ramp

Ammo track

LAYOUT AT A GLANCE

Overall size: 12 x 20 feet
Length of main line: 50 feet
Locale: Petersburg, Va.
Layout style: walkaround point-to-point
Layout height: 50"
Benchwork: sectional open-grid
Roadbed: cork on ½" plywood, cookie-cutter style

Track: code 83
Turnouts: No. 5, handlaid turnouts
Minimum radius: 18"
Maximum grade: none
Backdrop: .060" styrene sheet
Control system: wireless DCC, hand-thrown turnouts

Much of the turntable, located next to the enginehouse, was above ground with an elevated deck surrounding it. *U.S. National Archive*

sion occupies an extra foot in layout width, but requires an access aisle along the new extension.

The designed train length for this layout is one engine and nine cars. You'll have to create your own timetable, starting with two scheduled passenger trains. The rest of the freight trains can run as extras. The timetable can be adjusted once the layout has been put in operation and its operational quirks are discovered.

The general operating scheme has trains departing City Point yard for the various stations. Usually, the trains unloaded their goods immediately at the stations and returned. However, to add operational interest to the layout, setting out and picking up cars can be done with conventional waybills. A yard switcher works the spurs in City Point and assembles trains for the road crews.

Scenery should depict generally undeveloped countryside, with some farms and churches. There were large brick mills and structures in Petersburg, but they were far from the line and could be shown on the backdrop.

The army camps include the most numerous model "structures." The armies at the time used conical Sibley and rectangular tents. These can be made with scale lumber and tissue paper soaked in white glue.

The line of forts protecting City Point is shown on the National Archive maps. The tracks pass through these very close to Redoubt Five. Unfortunately, the maps do not give much detail about the nature of the redoubt's construction other than a profile of its earthworks. The map does not indicate the presence of artillery, but some pieces are undoubtedly present. These would be installed in fixed locations with firing platforms and revetments.

Near Meade Station the line was in direct line of sight of Confederate soldiers. The railroad erected earthworks along this stretch to shield trains from enemy fire. Part of this fortification is included in the plan.

The waterfront is probably the scenic highlight of the layout. Models should reflect the rough temporary style in which the real structures and wharves were built.

The plan intentionally includes ample room for ship models. Barges, tugboats, schooners, and steamships of all types from that era are possible. Many appropriate kits from the ship modeling side of the hobby can be used.

The plan assumes one is standing on the bluff looking down on the waterfront. The layout does not include the headland adjacent to the harbor. Thus Grant's log cabin headquarters, which was situated on this escarpment, is not included. However, the track leaving City Point leaves the flat harbor scene and pierces the bluff through a narrow ravine. At this location, you can model red clay dirt and scrub brush on the high ground above the waterfront.

The Loblolly Pine trees native to this area are difficult to model. The tried-and-true method of hand-carving wood trunks and adding branches from dried Caspia or other suitable weeds is a good, if tedious, approach. Fortunately not many are needed, as the soldiers quickly denuded the countryside of trees for firewood, lumber, and fortifications.

An officer and a civilian meet on a hill above the enginehouse and water tanks. The enginehouse is built on stilts to provide below-track inspection pits to service the locomotives. The locomotives were wood-burners, as evidenced by the rear view of the 4-4-0 at right. *U.S. National Archive*

Large numbers of scale soldier figures should be distributed about the layout, especially in the camps and stations. During the siege, the soldiers were relatively idle and bored when not performing their routine but dangerous guard and picketing tasks.

The arrival of a train always broke the monotony, and often meant packages from home or fresh food. Many photographs show soldiers relaxing near tracks without weapons and in a wide variety of military and civilian garb.

O scale plan

Although the O scale track plan doesn't permit the same level of prototype fidelity as the HO plan, the layout still conveys the feeling of and captures the operational potential of the line. The reduced track in the yard won't support operation of a prototypical number of trains. The engine terminal is shorter and reduced from three to two stalls. The main line and branches are also truncated but will permit realistic operations. If more room is available, the

Combat artist Alfred Waud sketched many scenes around the war zone, including this view of a typical station with telegraph lines, tents, and a lookout tower in the background. *U.S. National Archive*

line to Patrick Station can be extended similar to the HO plan. Although the plan is condensed, the increased detail and good running qualities inherent in larger-scale models should compensate.

Ship models are available in this scale to populate the harbor. A full-sized steam

sloop in this scale will approach three feet in length and would be a modeling challenge in its own right. Civil war military figures are not as plentiful in O scale. However, figures intended for narrow gauge layouts can be converted with a little bit of work.

CHAPTER 10

Wiscasset, Waterville & Farmington Railway

Northbound No. 3 pauses at the WW&F's Wiscasset depot. The Maine Central crossing is just south of the depot, and the Maine Central station is visible at right. *William Moneypenny photo; Gary Kohler collection*

Maine's two-foot gauge railroads have held a special attraction for modelers for a long time. The diminutive locomotives and cars, together with some rugged and impressive scenery, give these railroads a unique appeal. The Wiscasset, Waterville & Farmington, although not the best known of the Maine two-footers, has many features that lend it to modeling.

History

Maine's scenic and rugged coast features more than 3,500 miles of tidal shoreline (when all of the bays and inlets are added together), more than any other state except Alaska, Florida, and Louisiana. This massive coastal area provides numerous locations for deep-water, protected harbors. Among these is Wiscasset, on the shore of an expansive bay of the Sheepscot River. Located 13 miles from the ocean, Wiscasset was blessed with one of the finest natural harbors on the coast of northern New England. Founded in 1720, the town prospered in the years before the Revolution. The ancient forests in the vicinity of the Sheepscot River provided abundant oak, spruce, and white pine for export as well as for local shipyards and for the houses of the merchants and shipbuilders.

The 1790s and early 1800s were the time of Wiscasset's greatest affluence. There were shipyards, ropewalks, and sail lofts, and ships sailed from the harbor with lumber to trade for sugar, molasses, and rum from the West Indies and manufactured goods from Europe.

The Embargo Act of 1807 and the War of 1812 started a decline in trade that severely affected Wiscasset. Shipyards and trade houses moved to other towns, and the waterfront became nearly derelict.

Railroad to the rescue

Seeking to revive the town, the residents looked to the railroads. The Kennebec & Wiscasset Railroad received a charter in 1854, but no real progress was made.

In 1871 the Knox & Lincoln Railroad line from Woolworth (near Portland) passed through Wiscasset on its way to Rockland, providing a connection to Boston. Unfortunately, the line's construction costs, necessitated by numerous swamps, bridges, and an expensive ferry across the wide Kennebec River, impoverished the financiers, including the town of Wiscasset. To make matters worse, the Knox & Lincoln was not a success financially and Wiscasset was bankrupted. Several towns on the line actually voted in 1883 to abandon the line and sell the rails, but it survived. The Maine Central leased the railroad in 1891 and fully absorbed it in 1901, operating it as a branch.

Wiscasset, Waterville & Farmington, circa 1910

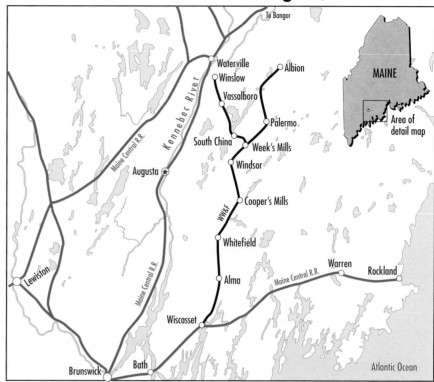

In the 1890s, a few wealthy locals formed the Wiscasset & Quebec Railroad, its name indicating its grandiose objective. In order to save on construction and operation costs, they decided the W&Q would be two-foot gauge. The founders hoped that the railroad would merge with one of the major lines, change to standard gauge, and create a western transcontinental connection, with Wiscasset the eastern end of a coast-to-coast system.

Construction began in June of 1894 in Wiscasset. The once-thriving waterfront area was nearly abandoned, so the railroad utilized it to construct its right of way. Several wharves were demolished, while others were rebuilt in anticipation of new railroad business. Much of the line in town was built on trestles, creating an extremely model-genic scene. Construction proceeded up the Sheepscot River to Whitefield and Weeks Mills. In November 1895, it reached Albion, 44 miles from Wiscasset. With debt mounting, the railroad halted construction. The hopes that Wiscasset would be the eastern terminus of a coast-to-coast system began to fade.

But there was business to be had in the Sheepscot Valley. The W&Q bought two brand-new Portland Company For-

ney locomotives, Nos. 2 and 3, in 1894. Revenue from operations provided enough income to cover operating expenses, but not to service the debt acquired in construction. As a result, the railroad went in and out of bankruptcy several times. In 1901, Leonard Atwood bought the W&Q and reorganized it as the Wiscasset, Waterville & Farmington Railroad.

Atwood attempted to revive the quest for a Western connection. He chartered two new companies to build an extension northwest with the objective of reaching the Sandy River Railroad, another two-foot railroad serving the mountainous forest area of north-central Maine. The Franklin Construction Company started building from Weeks Mills to the large town of Waterville. In the meantime, he proposed that the Franklin, Somerset & Kennebec would build southeast from the Sandy River Railroad in Farmington and meet the branch from Weeks Mills. If the FS&K could link with the WW&F, the two-foot lines would run from the mountains to the sea – not quite the coast-to-coast empire that the founders envisioned, but perhaps enough to make a successful enterprise. The FS&K never got started due to a lack of interest by the SRRR and resistance from the standard-gauge Maine Central, which did not want the competition from a statewide narrow gauge system.

Nonetheless, the FCC built from Weeks Mills to Winslow and started a bridge over the Kennebec River. The approaches were completed before the venture ran out of money. The bridge was never finished, but the line to Winslow was completed in 1902.

The WW&F absorbed the FCC line and its brand-new 28-ton Forney locomotive, which became WW&F No. 4.

The railroad designated the tracks to Winslow as the main line and relegated the Weeks Mills-to-Albion stretch to branchline status, possibly in anticipation of heavy traffic from the Woolen Mill in North Vassalboro and the large town of Waterville.

Business was decent but not as great as expected. The railroad hauled lumber from the woods, coal to the American Woolen Co., and mail, passengers, and miscellaneous goods for the valley.

The railroad wharf was a center of activity on the WW&F, as it housed

facilities for coal and passenger traffic. The coal originated in West Virginia or Pennsylvania and arrived via ship from Philadelphia or Baltimore. It was less expensive to ship the coal via the combined rail and water route, in spite of the manual transloading at the wharves, compared to all-rail routing. The wharf was also used to load lumber aboard ships for distant transport.

After 10 years of tight cash-flow operation, the railroad deteriorated from lack of maintenance. Still saddled with debt, it went into receivership in 1907.

Carson C. Peck, vice president of F.W. Woolworth, bought the WW&F for $93,000, a considerable bargain. He reorganized the WW&F Railroad into the WW&F Railway, paid off all debts, and began a series of improvements. Trestles were repaired, new machinery was added to the shops, and three locomotives were purchased. Number 5 was a Hinckley Forney from the Bridgton & Saco River Railroad. Numbers 6 and 7, brand-new Baldwins, were the largest locomotives the road ever had. Number 6 was a 2-6-2 Prairie built for freight, and No. 7 was a 2-4-4 assigned to passenger trains.

Several flatcars loaded with logs dominate this view looking north at the upper yard at Wiscasset. The car shop is at right. *William Moneypenny; Gary Kohler collection*

Forney No. 2 switches a car in front of Wiscasset Grain. A Bachmann On30 Forney could be kitbashed to represent this engine. *Gary Kohler collection*

WW&F in Wiscasset, Maine, circa 1932

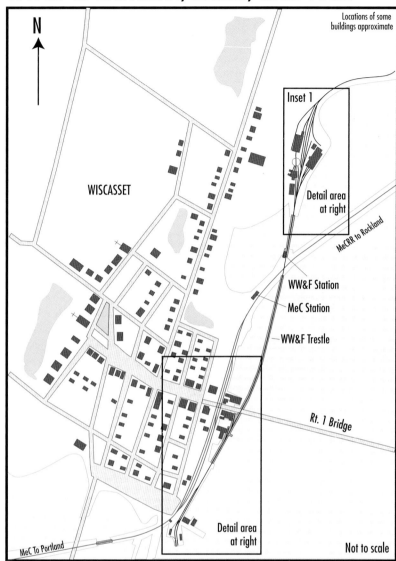

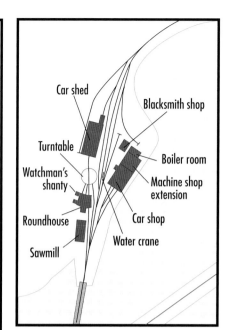

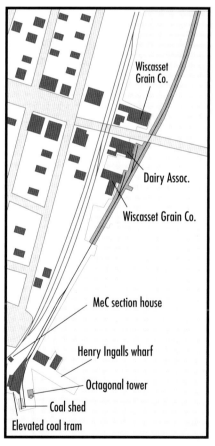

Coal was a major commodity on the WW&F as long as the woolen mill was a customer, but alas, it didn't last long. In 1914 , an interurban electric line, the Lewiston, Auburn & Waterville, reached Vassalboro and negotiated the coal contract for the woolen mill. Since Winslow itself had generated very little traffic, the portion from Winslow to North Vassalboro had already been removed in 1912. The loss of the woolen mill business resulted in most of the Winslow extension being pulled up by 1916, before the government took control of the railroads during World War I.

Business continued on the Albion line for the next 20 years. At the height of operations in 1921, the WW&F owned some 90 freight cars and six passenger cars, along with locomotives 1 through 7.

The railroad never made large profits, but managed to stay in the black. Agriculture from the Sheepscot Valley had become the primary business of the WW&F with potatoes, poultry, and lumber the main commodities. The railroad also had a mail contract and some passenger traffic.

Potato farmers left the Sheepscot Valley in the late 1920s, and as roads and highways improved, trucks began chipping away at the railroad's other business. The Great Depression slowed traffic even more, and in 1930 the WW&F again went bankrupt.

Frank Winter, a businessman with lumber interests in Palermo, then purchased the entire railroad. His main interest was to keep lumber shipments moving. Maintenance was not a priority, and he fired many employees to cut expenses.

An enginehouse fire in 1931 destroyed engines 6 and 7, and the remaining locomotives, Nos. 2, 3, and 4, were condemned the next year. To replace them, Winter bought the entire Kennebec Central Railroad in February 1933 just for its two engines. These Portland Forneys became Nos. 8 and 9.

The lack of maintenance on the track and roadbed took its toll, and four months later, in a span of just two days, both Forneys were irreparably damaged in accidents. When No. 8 jumped the track on June 15, 1933, the wreck ended rail operations on the WW&F.

Most of the mainline rail was ripped up for scrap in 1934 to satisfy unpaid bills, and in 1947 the engines and rolling stock were scrapped. Only one engine and a few cars survived the scrapper's torch. They are being brought back to life at the WW&F Railway museum in Alna, Maine.

Although the tale of the WW&F is full of false starts, bankruptcy, and eventual demise, the town of Wiscasset has survived to become a charming Maine seacoast tourist center. Several of the town's stately homes and municipal buildings have been preserved.

Modeling

There are several options for modeling the WW&F in HO or O Scale. Most two-foot modelers in HO use HOn2½, that is, HO scale (1:87) equipment on N gauge (9mm) track, which works out to 30" gauge. The appearance is still realistic, and it's a compromise that most HO two-foot-gauge modelers have learned to accept.

Over the years, WW&F engines have been released in HOn2½, usually as imported brass models. Although brass locomotives may be expensive, only a few will be needed. Further help for the budget comes from the short trains that the railroad typically ran. Most cars will have to be scratchbuilt or kitbashed, although some can be assembled from the few resin or laser-cut kits now available.

The wide availability of other HO items, such as scenery, structures, figures, vehicles, and detail parts, helps compensate for any scarcity of locomotives and rolling stock. The space-saving aspects of HOn2½ also make this choice of scale attractive, especially for a mid-sized space.

Options for track include handlaid or specially made HOn2½ track by Micro Engineering. Standard N scale could be

The depot at Vassalboro is typical of the small stations along the WW&F. *Gary Kohler collection*

Number 7 hauls several cars of logs past the Head Tide (Alna) depot around 1910. Note the elevated platform and the stub track at right. *Gary Kohler collection*

Wisscasset, Waterville & Farmington, HOn2½, circa 1910

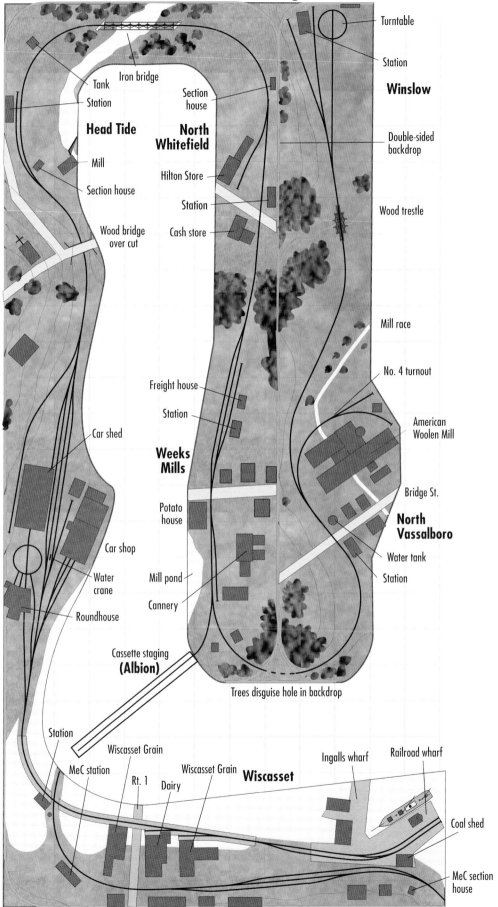

Turntable

Station

Winslow

Double-sided backdrop

Wood trestle

Tank

Station

Head Tide

Iron bridge

Section house

North Whitefield

Mill

Hilton Store

Section house

Station

Wood bridge over cut

Cash store

Mill race

No. 4 turnout

Freight house

Station

American Woolen Mill

Weeks Mills

Car shed

Bridge St.

Potato house

North Vassalboro

Car shop

Water tank

Water crane

Mill pond

Station

Roundhouse

Cannery

Cassette staging **(Albion)**

Trees disguise hole in backdrop

Station

Wiscasset Grain

Railroad wharf

Ingalls wharf

MeC station

Wiscasset Grain

Wiscasset

Rt. 1

Dairy

Coal shed

MeC section house

LAYOUT AT A GLANCE

Overall size: 12 x 24 feet
Length of main line: 74 feet
Locale: Sheepscot Valley, Maine
Layout style: point to point
Layout height: 50"
Benchwork: sectional open-grid
Roadbed: cork on ½" plywood
Track: code 55
Turnouts: Atlas No. 6
Minimum radius: 24"
Maximum grade: none
Backdrop: .060" styrene sheet
Control system: wireless DCC,
    hand-thrown turnouts

used, such as Atlas code 55 flextrack and turnouts with every other tie removed, but the ties will be noticeably too short. The WW&F Forneys and 2-6-2s require an 18" minimum radius and No. 6 turnouts. Turnouts with powered metal frogs are preferable because of the short wheelbases and limited coasting ability of the small locomotives.

The situation in O scale is a little better, with the option of On2½ (using HO gauge track) or true On2, with the track gauge at the correct two-foot width. Several brass locomotive models are available, and although relatively expensive, they are exquisitely and correctly detailed.

The second option, On2½, is a less-expensive alternative. Bachmann, which

calls it On30, for "30" gauge," makes several inexpensive, fine-running engines factory-equipped with Digital Command Control, that have helped make this a popular scale. Trains in On2½ can use HO track, but again, the ties will appear too short. Peco offers a line of track that is closer to scale with a rustic appearance. You can also handlay track to get the correct tie length and spacing.

Regardless of the scale you decide to model, the five-part book series by Gary Kohler and Chris McChesney, *Narrow Gauge in the Sheepscot Valley,* is a required reference for WW&F modelers. These volumes are loaded with photos, maps, and drawings of structures and equipment.

The HOn2½ layout

The HOn2½ plan utilizes the space-saving attributes of the scale to depict the year 1910 when the Winslow extension served as the main line. At this time the woolen mill was the main customer on the railroad.

The plan dedicates a good portion of its space to the town of Wiscasset and the signature waterfront scenes. The railroad

Engine No. 1 switches cars on the waterfront trestle while a Maine Central 2-6-0 waits with its train at the depot in the background. Several homes fill the hill behind the town. *Gary Kohler collection*

The octagonal tower building at left serves as the freight and passenger station on the railroad wharf. *Gary Kohler collection*

Wisscasset, Waterville & Farmington, On2½, circa 1932

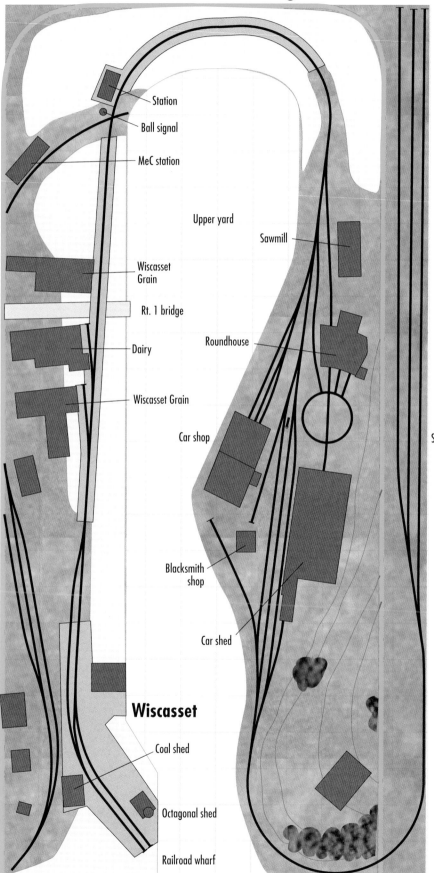

Station
Ball signal
MeC station
Upper yard
Sawmill
Wiscasset Grain
Rt. 1 bridge
Dairy
Roundhouse
Wiscasset Grain
Car shop
Blacksmith shop
Car shed
Wiscasset
Coal shed
Octagonal shed
Railroad wharf
Staging tracks

LAYOUT AT A GLANCE

Overall size: 12 x 24 feet
Length of main line: 50 feet
Locale: Sheepscot Valley, Maine
Layout style: point to point
Layout height: 50"
Benchwork: sectional open-grid
Roadbed: cork on ½" plywood, cookie-cutter
    style
Track: code 83
Turnouts: No. 6
Minimum radius: 30"
Maximum grade: none
Backdrop: .060" styrene sheet
Control system: wireless DCC,
    hand-thrown turnouts

wharf with its octagonal tower represents perhaps the quintessential New England dock scene. For added nautical appeal, you could include a model of a lumber or coal schooner.

Note that the waterfront area of Wiscasset evolved continuously over the years with changes to the wharves and structures. Consult Kohler's and McChesney's books, as well as other sources, for the most complete information.

The Maine Central tracks are depicted in truncated form, and the line is not actively operated in the layout. The WW&F and Maine Central did not interchange cars because of the difference in gauge. Cargo to be interchanged was manually transloaded from one railroad to the other. The Maine Central tracks were located slightly below the WW&F tracks so that the car floors could be on the same level. In operating the layout, cars on the Maine Central would be manually staged at the start of a session. The WW&F would treat these cars as another industry to switch.

The sidings on the trestle in front of the Wiscasset Grain Company buildings and the dairy are additional signature elements. These structures were built on pilings.

The dairy was built in 1913 by the Turner Centre Dairying Association, but is included in the plan for the additional switching. This business was once the largest commercial creamery in Maine and one of the three largest in New England, with dairies all over the state. It manufactured the first commercial ice cream in New England, and the association's founder, Edwin Leavitt Bradford, is credited with the invention of the celebrated Eskimo Pie, although many others around the country have made the same

The Manning Potato House in Weeks Mills had its own rail spur. *Gary Kohler collection*

The *Horatio*, a heavily laden lumber schooner, is tied up at the railroad wharf. Several boxcars are lined up in the background. *Gary Kohler collection*

claim. The dairy owned three narrow-gauge refrigerator cars lettered TDCA, and were likely the only non-WW&F cars on the line.

The WW&F station is located just north of the Maine Central junction, while the MEC station is just south of the junction. Semaphores and a ball signal, another signature New England item, controlled the diamond. Today the prototype WW&F is long gone, but tourist trains still use the former Maine Central tracks.

Another trestle leads to the upper yard. The upper yard was the principal maintenance facility on the WW&F. The track plan shows the arrangement present in 1910, so Winter's sawmill is absent.

The next modeled station on the line is Head Tide. The wooden bridge over the cut may not have been present in 1910, but it helps break up the scene from the upper yard. A classic New England-style mill, dam, and pond reside along the front fascia.

The iron bridge at Whitefield is another signature scene on the railroad. The delicate construction of this bridge will be a challenge to model. Heavy woods on each side of the bridge help isolate this scene.

Around the bend is North Whitefield, a small town with another signature scene, the Hilton Store. A model kit of this classic store is available.

Weeks Mills is an important location. The plan includes a passing siding and several spurs for a cannery, freight house, and station. In the prototype, some distance separated them; the plan compresses them into one location.

The wye at Weeks Mills is abbreviated to save space. The branch to Albion is simulated with cassette staging. This is designed so that a removable cassette can be used to receive and stage trains. Once or twice a session, a cassette can be installed to introduce or remove a train. Once accomplished, the cassette is stored under the layout, freeing up the aisle. Since the trains are short, a four-foot-long cassette should suffice.

The track for the Winslow extension passes through the double-sided backdrop. Trees hide the hole in the backdrop.

The tracks emerge on the far side in North Vassalboro. Here is the large

The brick woolen mill at North Vassalboro was the largest structure along the WW&F. The railroad primarily delivered coal to the mill. *Gary Kohler collection*

The corn cannery in Weeks Mills was another important business served by the WW&F. *Gary Kohler collection*

American Woolen Mills, with two stub sidings to switch. Several Walthers Front Street Warehouses could be kitbashed into a convincing model of this mill. The outbuildings and mill race would help make this an attractive scene.

The line continues to Winslow, where there's a small engine terminal to turn the engines and prepare them for the return trip. A wooded area and wooden trestle separate the Winslow and North Vassalboro scenes.

The On2½ plan

Despite the small size of On2½ equipment, there isn't enough space to model the line from Wiscasset to Winslow or Albion. Instead this plan focuses on modeling the Wiscasset waterfront and upper yard with a high degree of fidelity.

Staging tracks around the bend represent the rest of the line. As in the HOn2½ plan, the Maine Central tracks

are truncated. The large Wiscasset Grain and dairy structures hide the break in the MEC tracks.

The plan shows the upper yard as it was in 1932, with Winter's Sawmill being the primary addition compared to the earlier plan. The car shed has an extension, and there is an additional curved siding behind the blacksmith's shop.

The town and upper yard provide enough switching activity to keep two operators busy. The 1929 WW&F timetable shows two scheduled mixed trains: No. 8 departed Albion southbound at 5:50 a.m. and arrived in Wiscasset at 8:50 a.m., and No. 11 returned north at 1:30 p.m., arriving at Albion at 4:30 p.m. An extra train or two could be added to increase operational interest.

The fine-running On2½ locomotives chugging across the classic waterfront scene would make this an irresistible model railroad.

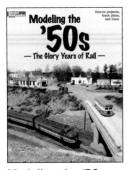
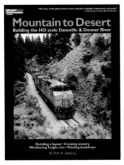
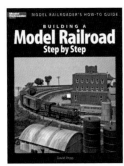
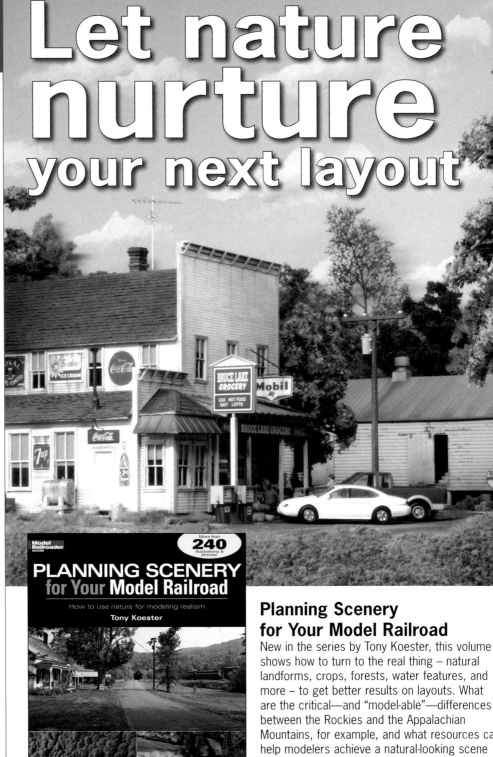